BIOFERTILIZERS IN INDIAN AGRICULTURE

Biofertilizers in Indian Agriculture

(An Annotated Bibliography, 1906-84)

L. L. Somani

CONCEPT PUBLISHING COMPANY
NEW DELHI-110015

ISBN 81-7022-029-7

First Published 1987

Published by
Naurang Rai
Conceept Publishing Company
H-13, Bali Nagar
NEW DELHI 110015 (India)

Printed at
New Gian Offset Printers
495 Sahzada Bagh
DELHI - 110035

Preface

Man began to cultivate land in an organized way for food grains around 8000 BC. Very soon, he learned that the same land cannot support the growth of plants endlessly and this led-him to think about the ways and means of improving the fertility of the soil. Man made fertilizers have come to stay in the present century for increasing the output of agricultural products and to meet the ever increasing demand of human population, which has been further accentuated by the limited availability of additional fertile farm land. The pathway of productivity improvement followed so far has however been accompanied by an exponential increase in the consumption of non-renewable forms of energy.

Industrial nitrogen for instance is heavily dependent on energy derived from fossil fuel which is getting depleted at a very fast rate. On the contrary, biological nitrogen fixation requiring approximately half the quantum of energy needed for industrial fixation is dependent on energy from ranewable resources such as products of photosynthesis and organic matter in soil. Therefore, it is essential for us to evolve and adopt a strategy of integrated nutrient supply by using a judicious combination of chemical fertilizers, organic manures and biofertilizers.

The term 'Biofertilizers' refers to preparations containing live or latent cells of efficient strains of nitrogen fixing, phosphate solubilizing or cellulotytic microorganisms used for application to seed, soil or composting areas with the objective of increasing the numbers of such microorganisms and accelerate certain micrlobiol processes to augment the extent of the availability of nutrients in a form which can be easily assimilated by plants. In a larger sense, the term may be used to include all organic resources (manures) for plant growth which are rendered in an available form for plant absorption through microorganisms or microorganism--plant associations or interactions. Thus, biofertilizers can make a significant contribution towards the development of strategices for productivity improvement which do not lead to an exponential rise in the consumption of non-renewable forms of energy.

The literature on this aspect is not only voluminous but also varied and widely dispersed in journals and reports. The objective of this book is to bring together the abstracts of published work into a small book form. The book will serve both biologist and agricultural scientists who are interested in alternative sources of energy on the farm through beneficial microorganisms.

An effort has been made in this book to provide an up-to-date account of fundamental as well as applied aspects of research in the field of biofertilizers in India. The publication of this book is timely when we are trying to evolve a new strategy for agricultural production based on integrated nutrient supply system.

I hope this book will be useful to research workers, teachers and planners in the field of biofertilizers as well as to other agricultural scientists, especially in the developing countries.

Needless to mention that I have drawn upon several standard authentic journals and abstract journals mainly in the libraries of SKN College of Agriculture: Jobner and IARI, New Delhi, and to all of them I express my acknowledgement with gratitude.

My acknowledgements are due to Dr. H.G. Singh, R.C. Mheta, Dr. K.K. Vyas, Dr. S.N. Saxena, Dr. S.S. Rathore, Dr. K.S. Singh, Dr. V.N. Pathak, Dr. H.N. Gour, Dr. K.L. Totawat, Dr. G.L. Jain, Dr. B.L. Baser, of Sukhadia University, Udaipur, who have helped me in the accomplishment of this work.

In the end, I crave the indulgence of my wary readers in the errors in which I must have fallen inadvertantly and solicit their valued suggestions towards improving the contents.

Jobner L.L. SOMANI

Contents

1.
ACHARYA, C.N. (1946) Relation between nitrogen conservation and quantity of humus obtained in manure preparation. Indian Fmg. 7(1) : 66-67.

Carbonaceous materials decompose faster if easily available nitrogenous material is added to bring the C/N ratio of the compost mixture up to 25-30. When the compost material undergoes decomposition the C/N ratio is lowered to 10. In good compost, the amount of mobile N from animal manure that has been converted into microbial tissues within the compost may be 30-50% of the total humus. If this mobile N is lost during preparation the quality of the manure is poor.

2.
ACHARYA, C.N. and JAIN, S.P. (1954) Nitrifiability of soil organic matter.

J. Indian Soc. Soil Sci. 2 : 43-48

Nitrification experiments carried out by incubating soils of different types in the laboratory under optimum conditions of moisture and temperature showed that: (1) the nitrification of soil nitrogen proceeded rapidly during the first three months, after which, the rate showed down and a stable maximum nitrate value was reached in about 6 months time; (2) among soils of different types, there was no direct relationship between the total nitrogen content of the soil and the quantity of nitrate formed, but, in general black cotton soils and alluvial soils nitrified readily to the extent of about 10% of the total soil nitrogen, whereas red soils and laterites nitrified more slowly to an extent of about 4% of the total soil nitrogen; (3) organic and other soils which possessed C/N ratios wider than 10 nitrified more slowly than soils with narrower C/N ratios; (4) the slowing down and ultimate stoppage of nitrification in the soil in about 6 months incubation in the laboratory, could be overcome by leaching out the soil with water, so as to remove the accumulated soluble products, and allowing the leached soil to nitrify again; the indications are that the accumulation of nitrate in the system or alterations in pH during nitrification are not the causes for stoppage of the nitrification process, but possibly water soluble bacterial toxins are formed during nitrification which inhibit the process.

3.
ACHARYA, C.N. and JAIN, S.P. (1955) Influence of the method of storage on the microbiological properties of soil samples.

J. Indian Soc. Soil Sci. 3 : 91-95

Storage of soil samples in bottles closed with cotton wool or cloth (aerobic storage) was compared against storage in bottles closed with rubber or glass stoppers (anaerobic storage) in regard to the effect of the treatments on the ammonifying, nitrifying and nitrogen fixing powers of the soil after 6 and 12 months of storage. In respect of ammonifying and nitrogen fixing powers, there was no appreciable difference between the two methods of storage, but the nitrifying power, as determined in Omeliansky and Winogradsky's media was much lower in the anaerobically stored samples, as compared to the aerobically stored samples.

The nitrifiability of soil organic matter, as measured by the quantity of nitrate formed by incubation of the soil under optimum conditions of moisture, temperature and aeration, was higher in the case of soil samples which had been previously kept stored under anaerobic conditions, as compared to soils which had been stored under aerobic conditions.

4.
ACHARYA, C.N. and JHA, J. (1954) Studies on the building up of soil fertility by phosphatic fertilization of legumes - influence of a legume rotation on the microbiological properties of the soil.

J. Indian Soil Soc. Sci. 2 : 21-28.

Field plots where berseem (Alexandrinum trifolium) had been grown in rotation with wheat and cowpea, showed higher microbial activity in respect of bacterial numbers, rate of CO_2 evolution, ammonifying and nitrifying powers than plots where no berseem had been grown. The microbial activity was higher in plots where berseem had been treated with superphosphate. The non-symbiotic nitrogen fixing power of the soil showed, however, no appreciable difference between the berseem grown and the non-berseem plots.

5.
AGARWAL, C.P. (1974) Heterocephalum aurantiacum, its occurrence in Indian soil and its cellulolytic activity.

Proc. Indian natn. Sci. Acad. 40 : 369-372.

Two isolates were obtained from Assam soil. Both decreased the strength of cambric cotton fabric.

6.
AGARWAL, C.P. (1975) A new cellulose-decomposing microorganism from Indian soils.

Proc. Indian natn. Sci. Acad. B 41 : 75-77.

Anixiella reticulata, hitherto unrecorded from soils of India, was recovered from soil of Tinsukia (Assam). The fungus was tested for cellulolytic activity by determining its ability to cause strength loss in cambric cotton strip and by its growth on different cellulosic materials. The isolate proved to be an effective cellulose decomposer.

7.
AGARWAL, S.C. (1980) A new species of Myrothecium from Indian alkaline soils.

Curr. Sci. 49 : 281-282

Myrothecium viride, a new species isolated from Usar soils (pH 8.0-9.5) collected from Aligarh, Uttar Pradesh is described.

8.
AGNIHOTHRUDU, V. (1953) Soil conditions and root diseases. VIII-Rhizosphere microflora of some of the important crop plants of south India. Proc. Indian Acad. Sci. 37B : 1-13.

Pot experiments with representative plants of the leguminosae, Pedalineae, Malvaceae and Gramineae grown on infected soil showed a significantly greater accumulation of fungi, bacteria and actinomycetes in the rhizosphere than in the control soils. The number of bacteria increased and of fungi decreased in the rhizosphere upto the time of flowering. Bacterial numbers continued to increase in the rhizosphere of non-flowering plants, and there was an increase in fungi and actinomycetes in the later stages of growth.

9.
AGNIHOTRI, V.P. (1963) Aspergilli from soil.

Naturwissenschaften 50 : 527.

Thirty species isolated and identified in Indian soils are listed. A fumigatus, A. flavus, A. niger. A. terreus and A. ustus were the most abundant.

10.
AGNIHOTRI, V.P. (1964) Studies on Aspergilli. XXV-Effect of foliar spray of urea on the Aspergilli of the rhizosphere of Triticum vulgare L. Pl. Soil. 20 : 364-370.

Sixteen species of Aspergillus were isolated from wheat rhizospheres. The largest number of species occurred near plants treated with urea (five sprayings); spraying increased the percentage of six species.

11.
AGRAWAL, S.C. (1971) The cellulolytic capacity of some forests litter fungi.
Phyron 28 : 169-175.

The classical cellulose enzyme of Pringsheim is composed of at least 2 entities designated C_1 and C_x. Celluloses produced in culture are designated C_x and those produced by organisms capable of hydrolysing or altering native cellulose are termed C_1. Tests with 15 fungi capable of acting on cellulosic sources when other soluble C sources are absent, showed the following categories; those with C_1 and C_x activity which are capable of acting on native cellulosic sources and carboxymethylcellulose (CMC); those with similar activity and producing in addition an enzyme without the need for any inducing substrate; those which produce C_x activity only and which require the presence of CMC as inducer; those producing C_x activity without needing an inducer. Rhizopus nigricans is exceptional in showing fairly poor activity (C_1 and C_x); it survives in the soil in the form of a utilizer living on the cellulosic decomposing products released by other fungi.

12.
AHMED, N. and JHA, K.K. (1977) Effect of inoculation with phosphate solubilising organisms on the yield and P uptake of gram.

J. Indian Soc. Soil Sci. 25 : 391-393

The effect of inoculation with two strains of P solubilisers on the yield and P uptake of gram was investigated in an acid red loam soil. The inoculation increased the yield and P uptake but it was more pronounced when FYM was mixed in the soil.

13.
AHMAD, N. and JHA, K.K. (1982) Effect of phosphate solubilizers on the dry matter yield of and phosphorus uptake by soybean.

J. Indian Soc. Soil Sci. 30: 105-106

Results show that the effectiveness of P solubilizers vary with the crops in the same soil. The solubilization of rock phosphate was found to be associated with the growth of microorganisms. Response due to inoculation with phosphate solubilizing organisms may not be universal and therefore, efficient strains should be identified for each crop and soil type for increasing crop yields.

14.
AJMAL, M.; AHMAD, A.; HASAN, M.Z. and NOMANI, A.A. (1980) Detrimental effects of pharmaceutical industrial waste on microorganisms. Water, Air, Soil Pollut. 13 : 347-352.

Pharmaceutical industrial waste was collected from Ghaziabad (Cooper Pharma Ltd.), analyzed, and its effects on microorganisms were studied. Detrimental effects, depending on the volume of waste applied, were observed on saprophytic and nitrifying bacteria.

15.
AJMAL, M.; NOMANI, A.A. and AHMED, A. (1984) Acute toxicity of chrome electroplanting wastes to microorganisms : adsorption of chromate and chromium (VI) on a mixture of clay and sand.

Water, Air Soil Pollut. 23 (2) : 119-127

Chrome electroplanting wastes were collected from two industrial sites and analyzed for colour, turbidity, pH, alkalinity, sulphate, chloride, N-ammonia, N-nitrate, N-nitrite, acid hydrolyzable P, dissolved oxygen, BOD, chemical oxygen demand. The effect of these wastes on saprophytic and nitrifying bacteria was studied with varying concentrations of wastes.

16.
AMMA, P.A.; AIYER, R.S. and SUBRAMONEY, N. (1966) Occurrence of blue green algae in the acid soils of Kerala.

Agric. Res. J. Kerala 4 : 141-143.

Nostoc and *Anabaena* are the predominant N-fixing organisms in these peaty rice soils. Liming favours the proliferation of these blue-green forms and suppresses the growth of green algae. The increased amounts of avilable N obtained with even low rates of liming may be due to the increased fixation of N by these powerful N fixers.

17.
ANANTHANARAYANA, R. and MITHYANTHA, M.S. (1970) Urease activitiy of the soils of M.R.S. (Main Research Station), Hebbal.

Mysore J. agric. Sci. 4 : 109-111.

The contribution of viable ureolytic micro-organisms to the urease activity in these soils is small. Activity of the enzyme is related to the moisture content and organic C.

18.
ANEJA, K.R. and MEHROTRA, R.S (1980). Comparative cellulolytic ability of microfungi inhabiting various types of litter. Proc. Indian natn. Sci. Acad. 46B: 566-571.

Relative cellulolytic ability of seventy three different litter-decomposing microfungi was measured *in vitro* by column clearance technique based on the clearing of acid swollen cellulose. The fungi are divided into four different groups on this basis. Maximum depth of clearing was shown by *Trichoderma harzianum* and species of *Fusaria*, *Aspergillii* and *Pencillia*. Pioneer colonizers of different types of litter had good cellulolytic ability.

19.
ANIL KUMAR, T.B. and CHAKRAVARTI, B.P. (1970) Changes in the rhizosphere microflora of maize seedlings by pretreatment of roots with hormones and urea.

Pl. Soil 33 : 697-684.

Gibberellic acid at 1 ppm stimulated rhizosphere organisms, while a concentration of 5 ppm suppressed populations of bacteria and actinomycetes; maleic hydrazide at 5 ppm and 1 ppm stimulated rhizosphere organisms (including fungi) and the bacterial and actinomycete floras, respectively. Urea stimulated the growth of fungi, bacteria and actinomycetes. The order of predominance of rhizosphere organisms was : fungi > bacteria > actinomycetes with 1-5 ppm gibberellic acid or 1 ppm maleic hydrazide, and bacteria > fungi > actinomycetes with 5 ppm maleic hydrazide, 0.1 M urea or untreated controls.

20.
APTE, R. and ISWARAN, V. (1974) Cultures of Rhizobium inoculants with those of Baijerinckia and Azotobacter. Proc. Indian natn. Sci. Acad. 40B : 482-485.

The survival of Rhizobium Japonicum cultured in peat with Beijerinckia indica and Azotobacter chroococcum was examined with special reference to nodulation and yield of soybean var. Clark in Delhi soil. There was better nodulation and increase in yield of the crop as compared with culture containing R. japonicum alone.

21.
APTE, R. and SHENDE, S.T. (1981) Studies on Azotobacter chroococcum. I. Morphological biochemical and physiological characteristics of Azotobacter chroococcum.

Zentbl. Bakt. Parasitkde. Abt. II 136 : 548-554.

Azotobacter strains were isolated from the rhizosphere soil of different agricultural crops. Their morphological and biochemical properties were studied in detail to identify efficent strains for the preparation of bio-fertilizers. All the organisms were Gram-negative. There were both pigmented and non-pigmented forms. They also differed in their main physiological attributes, such as nitrogen fixation and the synthesis of gibberellic acid (GA) and IAA-like compounds. Nitrogen fixation ranged from 1.50 mg to 9.50 mg/g of sugar consumed. No correlation was observed between pigmentation and nitrogen fixation. Except for the non-chromogenic strain (M-6), the strains all possessed differing capacity for producing indole compounds, ranging from 0.75 to 3.00/ug of IAA equivalents/ml of culture filtrates. The strains were all capable of synthesizing GA although they differed greatly in this respect; non-chromogenic strains such as M-6 and W-3 showed a poor capacity for GA synthesis.

22.
APTE, R. and SHENDE, S.T. (1981) Studies on Azotobacter chroococcum. II. Effect of Azotobacter chroococcum on germination of seeds of agricultural crops.

Zentbl. Bakt. Parasitkde. Abt. II. 136 : 555-559

The effect of strains of Azotobacter chroococcum on the germination of cotton, wheat and maize seeds was studied under three different conditions. The stimulatory effect of A. chroococcum varied with the strain and with germination conditions. Wheat and cotton seed germination was better in sterile petri-plates, but large seeds (maize) showed better germination in sterile sand cultures. Use of homologous strains of A. chroococcum, such as W-5 for wheat, C-2 for cotton and M-2 for seed inoculation of maize improved germination of these crops.

23.
APTE, R. and SHENDE, S.T. (1981) Studies on Azotobacter chroococcum. III. Establishment of Azotobacter chroococcum on the roots of crop plants.

Zentbl. Bakt. Parasitkde. Abt. II. 136 : 560-562

Fourteen strains of Azotobacter chroococcum were used for seed inoculation of wheat, maize and cotton to study their ability to establish in the root environment, under sterile sand culture conditions. Preferential growth on root segments of these plants was observed with various strains. The need for a consideration of plant-strain affinities during the selection of Azotobacter strains for bacterial inoculation was stressed.

24.
APTE, R. and SHENDE, S.T. (1981) Studiies on Azotobacter chroococcum IV. Seed bacterization with strains of Azotobacter chroococcum and their effect on crop yield.

Zentbl. Bakt. Parasitkde. Abt. II, 136 : 637-640

Seeds of wheat, maize and cotton were inoculated with different strains of Azotobacter chroococcum. The performance of the strains was evaluated in sowing experiments with and without a basal fertilizer dressing. Homologous selection of a pigmented strain (C-2 for cotton, M-2 for maize and W-5 for wheat) gave higher yield increases than non-pigmented strains.

25.
AROCKIANATHAN, T. and SARAF, C.S. (1984) Total dehydrogenase activity in chickpea cultivars in relation to date of sowing and row spacing. Golden Jubilee Seminar on Soil Resources and Productivity Management, Indian Society of Soil Science, New Delhi. 3 : 58-59.

Dehydrogenases have been taken as an index of total biological activiity of the soil. In order to ascertain the effect of the different cultivars the amount of dehydrogenase present in the rhizosphere soil was analysed at different stages of crop growth. The dehydrogenase activity was fairly higher in the second year than in 1981-82. Different dates of sowing showed an increasing trend, row spacings did not manifest very well with respect to total dehydrogenase at sowing as well as at harvest. The tall genotype HG 175 recorded marginal, increase in dehydrogenase than the semi-spreading (one Pusa 209). This indicated that the beneficial rhizobium strain belonging to chickpea group did not differ with respect to rhizosphere zone of different cultivars.

26.
ARORA, D. and GAUR, A.C. (1979) Microbial solubilization of different inorganic phosphates. Indian J. Exptl. Biol. 17 : 1258-1261.

A range of bacteria and fungi producing clear zones of solubilization of tricalcium phosphate on Pikovskaya's solid medium were tested for their ability to solubilize other phosphate sources in Pikovskaya's liquid medium. All the micro-organisms tested, including species of Pseudomonas , Bacillus, Escheridiia, Aspergillus and Penicillium, effectively solubilized tricalcium phosphate and hydroxyapatite. Rock phosphate was more resistant to solubilization, relatively small amounts being solubilized by most species. Fungi were more effective than bacteria. Among the fungi, Aspergillus awamori and A. niger were most effective, while Pseudomonas striata and Bacillus polymysca were the most effective bacteria.

27.
ARORA, D.K. and DWIVEDI, R.S. (1979) Rhizosphere fungi of Lens esculenta Moench antagonistic to Sclerotium rolfsii Sacc.

Soil Biol. Biochem. 11 : 563-566

The growth of Sclerotium rolfsii Sacc., a causal organism of root disease of Lens esculenta Moench. and that of several rhizosphere fungi from this host, was studied on agar plates staled by rhizosphere fungi of the host. Fungi, such as Trichoderma harzianum, Aspergillus flavus, A. niger Penicillium rubrum, Curvularis lunata and Fusarium poae, showed strong competitive ability and tolerance to the combined staling products of rhizosphere fungi. Growth of S. rolfsii on agar staled by these individual fungi was studied in detail. Emphasis was laid on interspecific competition, antagonistic activity, and colony and hyphal interactions between the pathogen and these fungi. T. harzianum mark- edly depressed the growth of S. rolfsii. Necrotrophic mycoparasitic activity of S. rolfsii towards A. flavus and A. niger was also observed.

28.
ARORA, D.K. and DWIVEDI, R.S. (1980) Mycoparasitism of Fusarium spp. on Rhizoctonia solani Kuhn. Pl. Soil 55 : 45-53.

Experiments on nutrient and staled agar were carried out to investigate the mycoparasitic activity of 10 fusaria against Rhizoctonia solani. Penetration and coiling by Fusarium oxysporum, F. semitectum and F. udum in and around R. solani hyphae were observed. Lysis of F. udum hyphae was observed inside R. solani hyphae showing the reverse of the normal direction of necrotrophic mycoparasitic relationships. The mycoparasitic activity of Fusarium spp. was reduced on agar plates staled by fungal growth products.

29.
ARORA, D.K. and UPADHYAY, R.K. (1978) Effect of fungal stating growth subs- tances on colony interaction.

Pl. Soil 49 : 685-690

The effect of staling growth substances on colony interaction between some dominant rhizosphere fungi and Sclerotium rolfsii was examined in vitro. Staling growth substances diffused in the agar created the change in the pattern of colony interaction. Measurement of growth inhibition in colony interaction was assessed with the help of a proposed model.

30.
ARORA, S.K. (1969) Blue green algae : a source of nitrogen and organic matter in paddy fields.

Il Riso 18 : 63-66.

The article stresses the role of blue-green algae in N-fixation in paddy fields. Nitrogen fixation was greater when the soil was inoculated with Anabaena.

31.
ARORA, S.K. (1972) Effect of basic slags and blue-green algae on nitrogen mobilisation in paddy field.

Il Riso 21 : 233-238.

The number of different types of amino acid and their total content and the amount of N fixed was greater in soil inoculated with Anabaena navicu- loides and Tolypothrix tenuis than in uninoculated soil, especially when P was added. The amount of amino acids was greater with Tolypothrix than with Anabaena.

32.
ARUNACHALAM, L. and VENKATESAN, G. (1984) Effect of

bio-fertilizer application on the yield of Sesamum.

Madras agric. J. 71 : 259-260

The sesamum seeds were treated with Azotobacter and Azospirillum. The results show increased yield of sesamum due to the inoculation of bio-fertilizers. There is good possibility of reducing the 50 per cent of the fertilizer nitrogen by the inclusion of bio-fertilizers without adversely affecting the yield.

33.
BAGCHI, D.K.; CHANDA, S. and BANERJEE, A. (1981) Effect of inoculation and phosphorus on dry yield and utilization of nitrogen by cowpea Indian J. Agric. chem. 14 : 163-166

A fodder variety of cowpea, T-2 was studied with four phosphorus doses (0, 60, 120 and 180 kg P_2O_5/ha) with or without inoculation of the seeds. Increase, due to inoculation, in dry yield was significant over no inoculation. Highest protein yield of 382.5 kg/ha was obtained with 120 kg P_2O_5/ ha. Nitrogen percentage in dry matter varied between 3.5 and 4.0. of the total nitrogen 48 to 50% was extracted as protein N in the first harvest whereas in the regr- owth cut it was 40-47%. Results indicate that this variety of cowpea with inoculation and phosphate fertilization upto 120 kg/ha, is promising for protein extraction.

34.
BAGYARAJ, D.J. and BHAT, B.N. (1970) Effect of heavy doses of ammonium sulphate on the microflora of soil and the rhizosphere of wheat.

Mysore J. agric. Sci. 4 : 460-465.

Increasing $(NH_4)_2SO_4$, application on a black soil increased the total bacterial population but decreased fungal and Azotobacter populations in the rhizosphere of Triticum aestivum. High rate of $(NH_4)_2SO_4$ increased the actinomycete population of the rhizosphere despite increasing their numbers in non-rhizosphere soil.

35.
BAGYARAJ, D.J. and CHABAPATHY, K. (1970) Studies on the rhizosphere microflora of sorghum as influenced by mixed cropping with four other plant species.

Mysore J. agric. Sci. 4 : 415-423.

Mixed cropping of sorghum with cotton, groundnut or chilli significantly incre- ased the bacterial, actinomycete and fungal populations of the rhizosphere. Mixed cropping with potatoes however gave small increases in the bacterial populations and reduced the fungal, actinomycete and *Azotobacter* populations. The maximum increase in rhizosphere population and the only signifiicant increase in *Azotobacter* populations occurred in mixed cropping with chilli.

36.
BAGYARAJ, D.J. and MANJUNATH, A. (1980) Response of crop plants VA mycor- rhizal inoculation in an unsterile Indian soil.

New Phytopath. 85 : 33-36

Cotton, cowpea and *Eleusine coracana* grown in an unsterile Indian soil low in available P were inoculated with the VA mycorrhizal fungus *Glomus fasci- culatus*. Inoculation increased mycorrhizal infection, shoot and root dry wt. and top P and Zn contents but did not effect Mn uptake.

37.
BAGYARAJ, J. and RANGASWAMI, G. (1966) On variations in rhizosphere effects of some crop plants.

Curr. Sci. 35 : 238-239.

The rhizosphere effect (rhizosphere/soil ratio) varied with crop species, depth of sampling, age of plants and groups of micro-organisms concerned. The rhizosphere effect progressively increased with depth from which the roots were taken. Microorganisms, especially bacteria including azotobacter, were more numerous on roots of legumes than of non-legumes. The rhizosphere effect was greater after 40 days than after 20 days of plant growth. The rhizosphere effect on bacteria was greatest in redgram (Cajanus cajan) followed by sunnhemp and cotton; the effect on actinomycetes was greatest in cotton and smallest in redgram; fungi were appreciable stimulated by gingelly (Sesamum indicum) followed by redgram, sunnhemp, ragi (Eleusine coracana) and cotton. Increased rhizosphere/soil ratios at greater depth were mainly due to the fewer number of organisms in soil at the lower layers.

38.
BAGYARAJ, J. and RANGASWAMI, G. (1967) Studies on the effect of foliar nutrient sprays on the rhizosphere microflora of Eleusine coracana.

Mysore J. agric. Sci. 1 : 176-186.

In pot experiments 1% w/v N, P and K solutions were sprayed on ragi plants at 15 day intervals, precautions being taken against soil drenching. P and K sprays suppressed the bacterial population; N, P and K all increased fungal population but decreased Azotobacter and N and K sprays increased actinomycetes.

39.
BAHADUR, K. and AGRAWAL, K.M. (1965) Influence of calcium ions on growth and nitrogen fixation in cultures of the nitrogen fixing microorganism Azoto- bacter chroococcum strain No.120/ 2-1B/2B at pH values 6.5, 7.0 and 7.5. Zbl. Bakt. 119 II 245-255.

A $CaCl_2$ concentration of 0.01g/100 ml culture medium was optimum at pH 7.5 for N fixation, and optimum at pH 6.5 for growth, of this strain.

40.
BAHADUR, K. and GAUR, N. (1979) Effect of ruthenium on nitrogen fixation by some nitrogen fixers.

Zentbl. Bakt. ParasitKde. Abt. II, 134 : 594-599

In culture medium, the nitrogen-fixing ability of three unidentified species of Azotobacter isolated from Allahabad soil was greatly increased by 25-75/um ruthenium chloride whereas sugar consumption remained more or less steady.

41.

BAHADUR, K. and GAUR, N. (1980) Effect of amino acids on microbial fixation of nitrogen. Zentbl. Bakt. ParasitKde. Abt. II 135 : 669-673.

The effect of 0.0025 and 0.005 M glutamic acid, proline and hydroxyproline on the nitrogen fixing ability of three nitrogen fixing microbes was investigated. Glutamic acid in the culture medium favoured N fixation by all three organisms. Proline also favoured N fixation but hydroxyproline had an inhibiting effect.

42.

BAHADUR, K. and JAISAL, U. (1973) Effect of uranium on microbial nitrogen fixation.

Zentbl. Bakt. Parasitkde Abt. II, 128 : 758-763.

Uranyl acetate at concentrations of 25/um to 50/um in the culture media increased N fixation and decreased C consumption by N-fixing bacteria.

43.

BAHADUR, K. and JYOTISHMATI, U. (1976) Effect of uranyl acetate on microbial fixation of nitrogen in presence of lanthanum sulphate.

Zentbl. Bakt. ParasitKde Abt. II, 131 : 157-162.

Uranyl acetate at 100/um increased N fixation by Azotobacter in a culture medium containing 25/um of lanthanum sulphate, but this increase was decreas- ed by higher concentrations of the sulphate.

44.
BAHADUR, K. and SINHA, K.K. (1965) Study of the influence of the sodium salts of fatty acids on the biological fixation of N_2. Zbl. Bakt. 119 Abt. II, 239-244.

The Na salts of different fatty acids were used as a source of C, in addition to mannitol, for various strains of Azotobacter. Phthalate, citrate, succinate fumarate and acetate were all utilized by the organisms, the order of utilization and resulting N fixation varying with strains tested.

45.
BAHADUR, K. and SORABJI, P. (1970) Ultraviolet irradiation on nitrogen fixing bacteria and determination of the inhibition and inhibition constant.

Zentbl. Bakt. ParasitKde. Abt II, 124 493-497.

Samples of bacterial cultures, grown on appropriate media, were given ultra- violet irradiation for 15-90 min. Effective N fixation gradually decreased with increasing intensity (time) of irradiation, due to rapid mutation of bacteria and ultimate lethal effects; N fixation was inversely related to bacterial inhibition.

46.

BAHADUR, K. and TRIPATHI, P. (1975) Effect of mannitol, sucrose, and glucose on bacterial nitrogen fixation.

Zentbl. Bakt. ParasitKde. Abt. II, 130 : 17-20.

Bacteria isolated from soil at Allahabad showed the highest in vitro N fixation efficieny in terms of N fixed per gram of carbon consumed in the presence of mannitol, followed glucose and finally sucrose.

47.
BAHADUR, K. and TRIPATHI, P. (1976) Nitrogen fixation by microbial cultures with sodium salt of organic acids as carbon source. Zentbl. Bakt. ParasitKde Abt. II, 131 : 375-376.

In culture of Azotobacter, Psedomonas, Rhizobium and Clostridium to which C was supplied as the Na salt of organic acids. N fixation and the amount of N fixed/g of C consumed decreased in the order : salicylate, benzoate' oxalate, citrate, acetate.

48.
BAHADUR, K. and TRIPATHI, P. (1978) Investigation of cerium and yttrium ions effects on microbial nitrogen fixation and determination of inhibition.

Zentbl. Bakt. ParasitKde Abt. II 133 : 628-631.

In culture medium, Y or Ce inhibited N fixation by Rhizobium, Azotobacter, Pseudomonas and Clostridium. C consumption increased with increasing concentr- ation of Ce.

49.
BAHADUR, K.; RANGANAYAKI, S. and MOHAN, C. (1981) Microbial fixation of nitrogen in presence of cobaltous chloride.

Zentbl. Bakt. ParasitKde Abt. II, 136 : 355-358

Five nitrogen fixing bacterial samples, isolated from Allahabad soil, showed an increase in nitrogen fixation at a particular concentration of cobaltous chloride in the culture medium; cobaltous chloride also slightly increased the rate of respiration as indicated by the carbon consumption of the micro-organisms.

50.
BAJAJ, I.C. and SUBBIAH, B.V. (1957) Nitrification in soils as affected by peri- odic leachings. Sci. Cult. 23 : 314-315.

Leaching at fortnigtly intervals for two months produced a higher NO_3 content in soil incubated at 35° and 33% saturated than was found in similarly treated soils, but unleached. Weekly leachings nearly doubled the NO_3 content.

51.
BAJPAI, P.D. and GUPTA, B.R. (1977) Studies on the exploitation of rhizobial potentiality for berseem (Trifolium alexandrinum

L.) under saline-alkali soils.

J. Indian Soc. Soil Sci. 25 : 62-68

In field experiments, inoculation with Rhizobium strains KB3, which was found to be comparatively salt tolerant, increased the yield of berseem, N per cent and N uptake significantly over uninoculated control under the ameliorative influence of gypsum applied to the soil. Pelleting of inoculated seeds with gypsum gave additional significant increases in green and dry matter and N uptake by berseem crop. In medium saline-alkali soil (pH 8.9 and ESP 25.4), seed inoculation was effective even in unamended soil, while in high saline-alkali soil (pH 9.4 and ESP 42.6), seed inoculation did not respond without amendment or without pelleting. Magnitude of inoculation effect either alone or with pelleting was relatively high at increased doses of gypsum applied to the soil than that at lower doses.

52.

BAJPAI, P.D. and GUPTA, B.R. (1979) Effect of salinity and alkalinity of soil on some important microbial activities.

J. Indian Soc. Soil Sci., 27 : 197-198.

The ammonifying power which ranged from 43 to 54.7 per cent was the highest in normal (50.8%) and lowest in saline-alkali soil (43.4%). The nitrifying power was considerably low in salt affected soils as compared to normal soils. The sulphur oxidising power, compared to normal (28.7%) was better in alkali soil (37.5%). Carbon mineralizing power determined as the capacity to oxidise carbon of FYM (semi-rotten) was observed maximum in normal soil (8.5%) and minimum in saline-slkali soil (1.7%).

53.

BAJPAI, P.D. and GUPTA, B.R. (1979) Effect of salinization, phosphorus fertilization and their interaction on the symbiosis between Rhizobium trifolii and berseem (Trifolium alexandrinum L.) J. Indian Soc. Soil Sci., 27 : 462-469.

The studies revealed that salinization of soil had a decisive adverse effect on the symbiosis showing thereby a significant reduction in biological yield, nodulation and nitrogen uptake of berseem crop even at the lowest level of salinity, i.e. ECe 5.0 mmhos/cm. About 50 and 90% yield reductions were observed at ECe of 10.0 and 17.0 mmhos/cm, respectively. The ECe 20.0 mmhos/cm proved too high for the crop to survive. It was also evident that highest salinity bearing crop growth (ECe 15.0 mmhos/cm/delayed nodulation and nodules formed were observed to be ineffective in appearance. Phosphorus fertilization, in contrast to salinization, significantly increased the number and dry weight of nodules, yield and nitrogen uptake up to 150kg P_2O_5/ha during 1971-72 and up to 100 kg P_2O_5/ha during 1972-73. Phosphorus application like salinization increased nitrogen concentration in plants. Salinity x phosphorus interaction with regard to all parameters of symbiosis was found significant showing thereby that responses to phosphorus application were maximum in normal soil and decreased with increasing salinity. At moderate salinity (ECe 5.0 mmhos/cm), 100 kg P_2O_5 gave the similar symbiotic responses as obtained at phosphorus dose of 50 kg P_2O_5/ha in normal soil. Plant mortality and ineffe- ctive symbiosis were specific to salinity and could not be corrected by phos- phorus addition.

54.

BAJPAI, P.D. and RAO, W.V.B.S. (1971) Phosphate solubilising bacteria. Part.I. Solubilisation of phosphate in liquid culture by selected bacteria as affected by different pH values.

Soil Sci. Pl. Nutr. 17 : 41-43.

In incubation studies with Bacillus megaterium var. phosphaticum isolated from phosphobacterin obtained from the USSR and Indian strains of B megaterium, B. circulans and Escherichia freundii solubilization of P supplied as $Ca_3(Po_4)_2$ took place over a wide range of pH values (5.2-8.5). B. megaterium var. phosphaticum was the most active form within this pH range, but the Indian strain of B. megaterium dissolved substantial amounts of PO_4 at pH 9. Each of the selected bacteria had a different optimum pH for maximum solubili- zation.

55.
BAJPAI, P.D. and RAO, W.V.B.S. (1971) Phosphate solubilising bacteria. Part II. Extracellular production of organic acids by selected bacteria solubilising insoluble phosphate.

Soil Sci. Pl. Nutr. 17 : 44-45.

Among the non-volatile organic acids produced as a result of the decomposition of glucose in Pikovskayas medium by Bacillus megaterium, B. circulans and Escherichia freundii were citric and lactic acids. These acids may be mainly responsible for the liberation of PO_4 in solution probably because of their chelating action.

56.
BAJPAI, P.D. and RAO, W.V.B.S. (1971) Phosphate solubilising bacteria. Part III. Soil inoculation with phosphorus solubilising bacteria.

Soil Sci. Pl. Nutr. 17 : 46-53.

In incubation studies with unsterilized soil fertilizer P was released in signi- ficant amounts by inoculated Bacillus megaterium var. phosphaticum in the treatment FYM + $(NH_4)_2SO_4$+ superphosphate. With sterilized soil, native P_2O_5 was released by this bacterium in the treatment FYM + $(NH_4)_2SO_4$ + aptite. With b. circulana, there was no significant increase in total available P_2O_5 in unsterilized soil except in the treatment $(NH_4)_2SO_4$ + superphosphate, but in sterilized soil available P_2O_5 increased with $(NH_4)_2SO_4$ + apatite treatment. In pot experiments with wheat the uptake of fertilizer P in treatments supplying $(NH_4)_2SO_4$ + superphosphate was increased by inoculation with B. megaterium or B. circulans. With cow pea both species of Bacillus appeared to be useful for increasing yields, total P_2O_5 uptake and uptake of fertilizer P and N. In a field trial on soil low in available P, P_2O_5 and N uptakes of berseem were significantly greater with the treatment FYM + rock P + phosphobacterin (supplying B megaterium var. phosphaticum) than with FYM and rock P. Fresh weight, dry weight and P_2O_5 uptake of berseem receiving superphosphate were greater when *B. circulans* was inoculated into the soil *B. circulans* may thus be useful as an inoculant in Indian soils low in organic matter as it utilizes even small amounts as an energy source and increase crop yields through the release of PO_4 from superphosphate and apatite.

57.

BAJPAI, P.D.; ARYA, R. and GUPTA, B.R. (1980) Comparative studies on de- composition pattern of some plant materials in two different soil conditions during winter season. *Indian J. agric. Res.*, 14 : 91-102.

Decomposition pattern of paddy straw, dhaincha and their mixture (1:1) was studied in 'vitro' under low temperature conditions of winter in normal and saline sodic soils. A burst in CO_2 evolution occurred after 3 days of incubation with all plant materials in both soil types. Decomposition in terms of CO_2 evolu- tion and nutrient release was highest in normal soil. Persistence of nitrite N was observed in saline-sodic soil. Apparent per cent losses of added plant materials viz. dhaincha + paddy straw; dhaincha and paddy straw were 38.9, 33.3 and 20.9 in normal soil and 28.1., 27.2 and 12.8 in saline-sodic soil, respectively. Maximum CO_2 evolution occurred after 90 days in mixture, dhaincha paddy straw (1 : 1) decomposed at the slowest rate. Dhaincha alone and together with paddy straw showed release of available N while paddy straw showed N retention. All the plant materials showed initial immobilization of P. In general, a slight decrease in pH occurred at later stages of decomposition. Soil organic matter built up in winter, showing wide C/N ratio.

58.

BAJPAI, P.D.; GUPTA, B.R. and RAM, B. (1978) Studies on survival of *Rhizo- bium leguminosarum* on two carriers as affected by moisture and temperature conditions.

Indian J. Agric. Res. 12 : 39-43.

Low grade Indian peat was superior to the admixture of FYM and tank silt as a carrier of *R. leguminosarum* for storage up to 2 wk, but thereafter both were equally good. A combination of 40% m.c. and a temperature of 30°C was the best for the proliferation of R leguminosarum, but 40% m.c. with 20° was better for storage. The safe storage period of R. leguminosarum in both carriers at 20, 30, 40, and 50°C was for up to 10, 8, 6, and 4 wk resp.

59.
BAJPAI, P.D.; GUPTA, B.R. and SINGH, CHANDRIKA (1974) A note on the effect of salinity on symbiosis between Rhizobium trifolii and Bersem (Trifolium alexandrinum).

J. Indian Soc. Sci. Soil. 22 : 375-376

A green house experiment in soil culture equiliberated at desired levels of salinity (ECe 4.5, 7.8, 12 and 15) was conducted and the seeds of berseem (Trifolium alexandrinum) inoculated with an already tested effective strains of Rhizobium trifolii were sown. Observations on qualitative characteristics and pattern of nodulation showed the effective symbiosis having bigger and pinkish nodules formed mostly on primary roots up to ECe 7.8. However, the proportion of effective nodules was lesser (58.2%) as compared to normal soil (92.5%). Germination, growth, number and dry weight of nodules, dry matter content of the plant and total nitrogen content considerably decreased with the increase in salinity.

60.
BAJPAI, P.D.; GUPTA, B.R. and SINGH, CHANDRIKA (1976) Some microbiological studies in salt-affected soils. II. Effect of gypsum and mulch on selected soil microflora and evolution of carbon dioxide. J. Indian Soc. Soil Sci. 24 : 48-52.

Population of total bacteria, azotobactor, nitrosomonas and nitrobactor, consider- ably increased due to application of gypsum and mulch either alone or in combination in saline-alkali soil but these increases were more pronounced when both the amendments were added to the salt-affected soil. Magnitude of microbial population was also affected by the stage of crop growth. Maximum bacterial population was noticed at earing stage while minimum at the harvesting, CO_2 evolution was greater at higher bacterial count and showed a positive correlation with the latter (r=0.9085).

61.

BAJPAI, P.D.; LEHRI, L.K. and PATHAK. A.N. (1974) Effect of seed inoculation with Rhizobium strains on the yield of leguminous crops. Proc. Indian natn. Sci Acad. 40B : 571-575.

Increased responses on the yield of leguminous crops have been obtained by the use of Rhizobium strains, isolated at U.P. Institute of Agricultural Sciences, Kanpur. Yield increases in crops of berseem (Trifolim alexandriunm), Lobia (Vigna catjang) and groundnut were to the extent of 74, 46 and 21 per cent respectively. Similarly Rhizobium strains of U.P. Institute of Agricultural Sciences tested for gram (Cicer arietinum), pea (Pisum sativum) and soybean (Glycine max) gave significantly increased yields of respective crops.

62.

BAL, D.V. (1927) The effect of varying concentrations of ammonia on the nitrifying power of the soil.
Agric. J. India 22 : 298-300.

A concentration upto 60 mg of ammonical N per 100 g of soil does not materially affect the process of nitrification, but concentrations over 100 mg of NH_4-N are definitely injurious to the process of soil nitrification.

63.

BAL, D.V. and KRISHNAMURTY, R.S. (1938) Effect of sunlight on the nitrification of ammonium sulphate and oil-cake in the soil.

Proc. 25th Indian Sci. Cong. 225.

Absence of nitrification on exposure of soil to sunlight is not due to the death of the soil nitrifying organisms but to their temporary inactivity as a result of the unfavourable effects produced by sunlight. When soils previously exposed to sunlight are subsequently incubated in the dark the organisms recover and function normally.

64.
BALASUBRAMANIAN, A. and RANGASWAMI, G. (1973) Influence of foliar application of chemicals on the qualitative changes in bacterial and fungal popu- lation in the rhizospheres of Sorghum vulgare and Crotolaria juncea.

Madras agric. J. 60 : 218-224.

In pot experiments on red sandy loam soil, foliar spraying with $NaNO_3$ or Na_2HPO_4 reduced the percentage of gram-negative bacteria in the rhizosphere of sorghum and sunnhemp. Spraying with 2, 4-D and Dithane Z-78 increased the number of gram negative bacteria. $NaNO_3$ and Na_2HPO_4 promoted spore- formers and Dithane Z-78 and especially 2, 4-D reduced their number. $NaNO_3$ and 2,4-D both increased NO_3 reducers and NH_3 producers. Na_2PHO_4 and Dithane Z-78 decreased the number of NO_3 reducing bacteria and in sorghum suppressed NH_3 producers. Spraying with $NaNO_3$ stimulated bacteria requiring amino acids whereas spraying with Na_2HPO_4 suppressed them. $NaNO_3$ and 2, 4-D sprays stimulated Aspergillus but suppressed Penicillium. The presence of Helminthosporium and Fusarium in the rhizospher of sorqhum receiving these sprays when normally they were absent suggested that an increased availability of root exudates may have overcome the normal inhibiting effect of the roots on these pathogens.

65.
BALASUBRAMANIAN, A. and RANGASWAMI, G. (1973) Influence of foliar appli- cation of chemicals on the root exudations and rhizosphere microflora of Sor- ghum vulgare and Crotalaria Juncea.

Folia Microbiol. 18 : 492-498.

Sodium nitrate spray decreased amino acid exudation in sorghum but increased it in sunnhemp. Disodium hydrogen phosphate caused qualitative changes in sugars and reduced amino acid exudations. 2,4-D spray increased exudation of sugars and enhanced that of amino acids in both species. Dithane Z-78 reduced amino acid exudations and caused qualitative changes in sugars. Increases in bacterial population by 2,4-D in actinomycetes by 2,4-D, disodium hydrogen phosphate and sodium nitrate sprays, and in fungal populations by 2,4-D, and sodium nitrate were observed in sorghum rhizosphere. In sunnhemp rhizosphere 2,4-D and sodium nitrate sprays increased the bacterial, actinomycetes and fungal populations.

66.
BALASUBRAMANIAN, A. and RANGASWMI, G. (1978) Influence of seed and root exudations on the rhizosphere effect in Sorghum Vulgare and Crotolaria juncea.

Folia Microbiol. 23 : 481-488.

Although quantitative differences were observed between the two plant species in some morphological and physiological groups of bacteria on the seeds and rhizosphere, there were similarities in the occurrence of certain physiological and nutritional groups of bacteria on the seed as well as on the rhizosphere.
The occurrence of some genera of fungi on the seeds and in the rhizosphere indicated the influence of both seeds and root exudations of amino acids and sugars correlated with a concomitant increase in microbial activity in the rhizosphere on the 15th day of growth in both plant species, and there was a delayed effect of the exuded chemicals on the rhizosphere microflora after 30 days of plant growth.

67.
BALASUBRAMANIAN, A. and SIDDARAMAPPA, R. (1974) Effect of simazine application on certain microbiological and chemical properties of a red sandy loam soil.

Mysore J. agric. Sci. 8 : 214-219.

Simazine applied at the normal rate and ten times this rate led to decreases in bacteria and increases in actinomycetes, with no clear effect on fungi. Irregular changes in soil amylase and invertase activities were recorded. CO_2 release was reduced by the herbicide, while, at the normal rate, nitrification was stimulated.

68.
BALASUBRAMANIAN, A. ; BAGYARA, D.J. and RANGASWAMI, G. (1970) Studies on the influence of foliar application of chemicals on the microflora and certain enzyme activities in the rhizosphere of _Eleusine coracana_ Gaertn.

Pl. Soil 32 : 198-206.

Foliar application of $(NH_4)_2SO_4$, Na_2HPO_4 and KCl increased the activity of enzymes hydrolysing sucrose and starch in the rhizosphere of finger millet. Application of Dimecron 100 and Blastin significantly reduced enzyme activity during the early stages of growth.

69.
BALASUBRAMANIAN, A.; SHANTARAM, M.V.; SARDESHPANDE, J.S. and RANGASWAMI, G. (1972) Studies on certain microbiological properties of the old permanent manurial plots at Coimbatore, Tamil Nadu.

Madras agric. J. 59 : 443-448.

The microflora of surface soil samples from variously fertilized plots were studied. P, K and farmyard manure favoured bacterial growth, fungal growth and N actinomycete growth.

70.
BALASUBRAMANIAN, A.; SHANTARAM, M.V.; SIDDARAMAPPA, R.; EMMIMATH, V.S. and RANGASWAMI, G. (1974) Microbiological decomposition of organic matter in an alkali soil of Karnataka.

Mysore J. agric. Sci. 8 : 103-110.

Samples of an alkaline soil (pH 8.8, ESP 18.6, SAR 18.0) were incubated for 30 weeks with 0.5 and 2% organic C as FYM or jowar stalks. CO_2 evolution was greatest from soil containing 2% C as jowar stalk, this treatment also had larger fungal and bacterial populations than the other treatments. Numbers of actinomycetes were greatest in samples containing FYM (2% C). NH_3-N decreased and nitrification rate and release of available P increased during the incubation. Organic C, total N, humic and fulvic acids and CEC increased in samples containing 2% C as FYM or jowar stalk.

71.
BALASUBRAMANIAN, M. (1971) Root-knot nematodes and bacterial nodulation in soybean.

Curr. Sci. 3 : 69-70.

Meloidogyne javanica, *M. hapala* and *M. incognita* larvae inoculated into pots of soya-beans all reduced nodulation and root-hair production. *M. javanica* caused the greatest reduction and the nodule index decreased and root-knot index increased most with an increase in larval inoculum level from 10 to 100.

72.
BALASUBRAMANYA, R.H. and PATIL, R.B. (1980) A simple and economic medium for mass production of *Rhizobium* and *Azotobacter*.

Indian J. exp. Biol. 18 : 1204-1205.

Mannitol, the sole source of carbon in the media employed to mass produce *Rhizobium* and *Azotobacter*, was replaced by tamarind kernel powder after acid hydrolysis. The sugars released after neutralization supported growth of these organisms which compared very well with that on mannitol.

73.
BALASUNDARAM, V.R. (1974) Nodulation of soyabean under Indian conditions. Proc. Indian natn. Sci. Acad. 40B : 512-515.

Various methods adopted in overcoming the failure of nodulation of soyabean due to high soil temperature are presented.

74.
BALASUNDARAM, V.R. and SEN, A. (1974) Effect of climate on growth and nodulation of soyabean *Glycine max* (L.) Merrill.

Proc. Indian natn. Sci. Acad. 40B : 624-628.

Soyabean var. "Bragg" was grown in pots under natural conditions of Delhi for two months throughout one year. Three strains of *Rhizobium japonicum* belonging to three serogroups were used for inoculation. Climatic factors like temperatures, rainfall, sunshine hours and humidity were correlated with nodule number, dry weight of nodules and dry weight of plant tops at the end of the growth period. It was found that temperature and rainfall were found to have positive effect and sunshine hours to have negative effect on the performance of the strains. It is also observed that differently to the above factors.

75.
BALASUNDARAM, V.R. and SEN, A. (1977) Grouping of rhizobial strains-a method based on symbiotic characteristics.

Zentbl. Bakt. ParasitKde. Abt. II 132 : 616-622.

In a pot experiment, the Mahalanobis D^2 statistical method was used to classify twenty strains of *Rhizobium japonicum* according to their effects on seven plant characteristics of two varieties of soya-bean. Eight groups of *R. japonicum* were established, of which two differed greatly from the others.

76.
BALASUNDARAM, V.R. and SEN, A (1977) Stepwise selection of efficient rhizobial cultures through cultural characteristics.

Zentbl. Bakt. ParasitKde Abt. II. 132 : 623-627.

In the laboratory, nodulation in one variety of soya-bean by twenty strains of *Rhizobium japonicum* was correlated with the dehydrogenese activity and growth of the strains in a medium containing soil extract, whereas nodulation in a second soya-bean variety was correlated with the growth of *R japon-*

icum in media containing asparagine or tryptophane. The nitrogen content of the shoot of the first soya-bean variety was positively correlated with glucose consumption and the growth of R japonicum strains in Elkan's medium, whereas that of the second soya-bean variety was correlated with the growth of R japonicum strains in a medium containing aspartic acid.

77.
BALASUNDARAM, V.R. and SUBBA RAO, N.S. (1974) Researches on the establi- shment of soyabean in India through indigenous inoculants. Proc. Indian natn. Sci. Acad. 40B : 636-640.

Methods adopted in isolating, screening and selecting efficient strains of Rhizobium japonicum are outlined. Using Indian peat as the carrier, their performance is compared with the imported cultures at different agro-climatic conditions in the country. Experimental results of three years show that it is possible to establish soyabeans in India through indigenously available cultures. The impor- tance of composite cultures is stressed. Future strategies in improving the quality of the Rhizobium cultures to suit different agro-climatic conditions and varieties of soyabeans are also discussed.

78.
BALKRISHNAN, S. and RAJ, J.S. (1971) Effect of T.M.V. and certain chemicals on the rhizosphere microflora of tobacco.

Agric. Res. J. Kerala 8 : 114-120.

In pot experiments 65-day old tobacco plants growing in a mixture of equal parts of FYM, sand and soil were inoculated with tobacco mosaic virus 24 hr after being sprayed with aqueous solutions of 0.83 M urea, 10 ppm 2,4-D and 0.0012 M terramycin. Total bacterial, actinomycete and fungal counts in the rhizosphere were made at 6, 16 and 26 days after spraying and for 130 days in unplanted control soil and in the rhizosphere of untreated plants. In the rhizosphere of un-treated plants the numbers of bacteria, actinomycetes and fungi increased up to day 110 and then decreased. Infection with tobacco mosaic virus increased rhizosphere populations.

Urea had a slightly inhibitory effect on rhizosphere populations of uninfected plants but a stimulating effect on the rhizosphere of both healthy and infected plants and 2,4-D had stimulatory effect on both groups of plants.

79.
BANDYOPADHYAY, B.K. and BANDYOPADHYA, A.K. (1983) Effect of salinity on mineralisation and immobilization of nitrogen in a coastal saline soil of West Bengal. Indian Agric. 27 : 41-50.

The rates of mineralization and immobilization of nitrogen in soil were consider- ably reduced by soil salinity of 10 mmhos/cm and above. When rice straw was added @ 20 t/ha, about 51 per cent of mineral-N (NH_4-NO_3-N) in soil was immobilized at the end of the first week, and at the 40 t/ha rate about 92 per cent was immobilized.

80.
BANERJEE, S. K. (1974) Homeostatic control in nodulation. Proc. Indian natn. Sci. Acad. 40 B : 499-506.

Seeds of two varieties of soyabean were treated with different doses of X-rays, X_1 seedlings and distinct variants of X_3 population were inoculated with four serologically distinct strains of *Rhizobium japonicum* both individually and mixed in equal proportion.

In general, some proportionality was found between the induced variability in plant characters (excluding nodulation) and the X-ray doses employed. The induced variation in nodule number and total nodule volume could be represented by a normal unimodal curve plotted against the increasing doses of X-rays showing that there exists an autonomous control which tends to conserve the normal limit of nodular character even when subjected to X-irradiation.

Changes induced by X-rays in the host genotype did not produce any significant change in the number of respective nodule sites of individual competing strains in a mixed inoculum. The existence of an autonomous control in nodulation is thus further supported.

81.
BANIK, S. and DEY, B.K. (1981) Phosphate-solubilizing microorganisms of a lateritic soil. I. Solubilization of inorganic phosphates and production of organic acids by microorganisms, isolated in sucrose calcium phosphate agar plates. II. Effect of some tricalcium phosphate-solubilizing microorganisms on available phosphorus content of the soil. III. Effect of inoculation of some tricalcium phosphate-solubilizing microorganisms on available phosphorus content of rhizosphere soils of rice (*Oryza sativa* L. cv. IR-20) plants and their uptake of phosphorus.

Zentbl. Bakt. Parasitkde. Abt. II, 136 : 478-486.

Incubation tests showed that the available P content (Bray and Kurtz) of a lateritic soil treated with rock phosphate, manure and rice straw was increased slightly by inoculation with (a) *Bacillus firmus,* (b) a mixed culture of *B. circulans* and *B. subtilis,* (c) *Streptomyces* sp., (d) *Aspergillus* sp. and (e) *A. niger* isolated from the same soil and shown to be efficient in solubilizing inorganic P. P uptake and dry weight of pot-grown rice plants were increased significantly only by a mixed inoculant containing three *Bacillus* species.

82.
BANIK, S. and DEY, B.K. (1983) Alluvial soil microorganisms capable of utilizing insoluble aluminium phospate as a sole source of phosphorus.

Zentb. Mikrobiol. 138 : 437-442.

All cultures except *Bacillus* sp. AAB_5, *Streptomyces* sp. AAS_5, *Chaetomium* sp. AAF_5, Solubilized $Ca_3(PO_4)_2$ with greater efficiency than $AlPO_4$. *Bacillus coagulans (B. 7659)* AAB_2 showed highest efficiency in solubilizing $Ca_3(PO_4)_2$ followed by *Penicillium* sp. AAF_6. *Bacillus* sp. AAB_1, *B. subtilis* (B-7660), *B. megatherium* (B-7661) AAB_4 *Streptomyces* sp. AAS_4 and *Penicillium* sp. AAF_6 were unable to solubilize $AlPO_4$ in detectable amounts exceeding their nutritional demand. Production of oxalic acid was found to be less efficient in $AlPO_4$ solubilization. The amount of free organic acids present in the growth medium during solubilization was not always indicative of phosphate solubilization.

83.
BARDIYA, M.C.; and GAUR, A.C. (1970) Effect of some chlorinated hydrocarbon insecticides on nitrification in soil.

Zentbl. Bakt. ParasitKde Abt. II, 124 : 552-555.

Inhibition of nitrification of $(NH_4)_2SO_4$ in a sandy loam (in the laboratory) by 25 ppm or higher concentrations of insecticides was temporary (duration given in brackets) and followed the order; lindane (3 wk) > dieldrin (2 wk) > aldrin (1 wk). Decrease in NO_3 levels was accompanied by NO_2. accumulation. Results indicate that the insecticides tested do not harm nitrification when applied at the recommended field rates.

84.
BARDIYA, M.C. and GAUR, A.C. (1974) Isolation and screening of miicro-organisms dissolving low-grade rock phosphate.

Folia Microbiol. 19 : 386-389.

Several yeasts, fungi and bacteria isolated from the rhizosphere of leguminous crops and soils of a rock, phosphate deposit area were found to dissolve low-grade rock phosphate. *Schwanniomyces occidentalis*, *Aspergillus awamori* and *Penicillium digitatum* were particularly effective in this respect. Gram-negative motile rods were more effective than gram-negative non-motile rods. The most effective bacteria were *Pseudomonas striata*. All the micro-organisms acidified the liquid medium, but there was no relationship between the rock phosphate dissolved and the decrease in pH of the culture broth.

85.
BARDIYA, M.C.; NARULA, N. and VYAS, S.R. (1972) Effect of inoculation of *Thibacillus* on the lucerne crop (*Medicago sativa* L.) grown in alkali soils. J. Res HAU, 2 : 286-290.

In two separate pot culture experiments, highest yield of lucerne crop was obtained in the reclamation method of applying elemental sulphur and inoculating *Thiobacillus* as compared to gypsum amendment and half gypsum and half sulphur application. The inoculation of *T. novellus* (T 16) gave significantly higher yield as compared to *T. thiooxidans* (T 6). In Karnal soil, the yield in gypsum treatment was significantly higher over the sulphur alone (without inoculation) but in Rajpura soil yield was, of course, but it was statistically not significant and, furthermore, the inoculation of *Thiobacillus* with gypsum did not increase the yield.

86.
BARDIYA, M.C.; NARULA, N. and VYAS, S.R (1972) Reclamatson of saline alkali soil by the application of sulphur and inoculation of *Thiobacillus*. I-Effect on physico-chemical properties of soil. *J. Res. HAU*, 2 : 218-221.

The pH was found to reduce significantly under both sterilised and unsterilised conditions. The total soluble salts at any stage of incubation and under both the conditions remained higher in the treatment where both sulphur and *Thiobacillus thiooxidans* had been added, as compared to all other treatments. While soluble calcium increased steadily upto the end of incubation in sterilised series, in unsterilised series it increased upto 30th day of incubation and then gradually decreased. The maximum amount of soluble calcium was recorded again in soil + sulphur + *Thiobacillus* treatment.

87.
BAROOAH, P.P. and SEN, A. (1959) Studies on Beijerinckea from some acid soils in India.

Indian J. agric. Sci. 29 : 36-51.

Beijerinckia spp. occurred in 40% of the 50 different acid soils investigated, mainly different strains of *B. indicum* and *B. indicum* var. *alba*. The N-fixing capacity of the strains varied from 5.97 to 11.57 mg/g glucose during a period of one month.

88.
BARTHAKUR, H.P.; AZIZ, T. and WATANABE, I. (1983) Effect of rice straw application on the activity of algae in rice field. J. Indian Soc. Soil Sci; 31 : 146-147.

Surface applied rice straw with wide C : N ratio stimulated the growth of various microorganisms resulting in more CO_2 production and simultaneous depletion of O_2 and available nitrogen from the soil. Under such conditions too the growth of blue-green algae was stimulated in the soil. At the same time surface applied rice straw appeared to have provided physical support for better growth of blue-green algae in submerged soil.

89.
BARUAH, M. and MISHRA, R.R. (1983) Carbon-dioxide evolution and microbial population in flooded and upland rice (Oryza sativa L.) field soils of North East India. J. Soil Biol. Ecol. 3 : 69-76.

The results demonstrate positive relation between fungal population, fungal biomass and CO_2 evolution in flooded soil. Higher fungal and bacterial populations, CO_2 evolution, organic carbon, total N and P were constantly recorded in flooded soils. Fungal population was predominantly responsible for soil respiration.

90.
BASAK, M.K. and GOYAL, S.K. (1980) Studies on the biology of tree legumes Rhizobium symbiosis : Nodulation pattern and cross inoculation trials with the tree legumes and cultivated legumes.

Ann. Arid Zone 19 : 427-431.

Inoculation of Vigna radiata and V. mungo seeds with Rhizobium strains from some tree legumes increased the nodulation and seed yields compared with inoculation with the specific Rhizobium strains and uninoculated controls. Rhizobial strains of the tree legumes tolerated a temperature of 35-40°C while the specific rhizobia had a temperature optima of 30°C.

91.
BATRA, L. and BHARDWAJ, K.K.R. (198) Growth, nodulation and nitrogen fixation in legumes as affected by soil sodium. J. Indian Soc. Soil Sci. 29 : 530-536.

Growth, nodulation and nitrogen fixation in ten cultivated legumes were studied in an alkaline soil in the ESP range of 11 to 70. At higher ranges of sodicity there was marked reduction in dry matter yield, number of nodules and nitrogen content of all the legumes except dhaincha which was found to be distinctly tolerant to high exchangeable sodium. Only dhaincha and Indian clover could grow and nodulate at ESP 70. In other plant species, substantial growth and nodulation occurred only up to ESP 34. The number of nodules declined more sharply than the dry matter yield. In the initial period despite plant growth, nodulation was found to be lacking at the highest level of ESP. Subsequent appearance of nodules showed inhibition of legume-rhizobia symbiosis due to excess exchangeable sodium.

92.
BEHERA, N. and PATNAIK, R.S. (1980) Rhizosphere microflora of intact and defoliated plants of Dolichos biflorus L.

Acta Bot. Indica 8 : 105-109.

Number of bacteria and fungi in the rhizosphere of D. biflorus were reduced somewhat by repeated defoliation that left only two young leaves at the top of each plant.

93.
BERGTHEIL, C. (1906) The study of fermentation as applied to agriculture.

Agric. J. India 1 : 68-75.

Soil microorganisms are known to play important role in rendering soluble and available for plant food from insoluble materials. Legume bacteria obtain their N from some sources other than the soil. Their root residue and other parts left in the soil after the harvest of the crop add to the N status of the soil. Soil microbes also ferment other organic residue and release the nutrients held in them.

94.
BERGTHEIL, C. (1907) Soil inoculation for indigo. Agric. J. India 2 : 291-292.

Experiments on the inoculation of Behar soils with cultures of the indigo nodule bacteria showed no benefit. It appears that these soils are sufficiently well provided with the necessary organisms, and that these organisms are in as active a condition as can be obtained by artificial culture.

95.
BHADUR, K. and LAL AGRAWAL, K.M. (1966) Influence of calcium ions on growth and nitrogen fixation in cultures of the nitrogen fixing Azotobacter chroococcum strain No. 120/2 /1B/2A at pH values 6.5, 7.0 and 7.5.

Zentbl. Bakt. ParasitKde Abt. II, 120 : 703-711.

Fixation of N increased with increasing pH to up to mg per g of C consumed in the presence of 0.001 2/100 ml $CaCl_2$ in the culture medium. Growth of the organism was little affected by pH, but was greatest at pH 6.5 in the presence of $CaCl_2$.

96.
BHADUR, K. and TRIPATHI, P. (1978) Investigation of inorganic and organic nitrogenous compounds on microbial nitrogen fixation with respect to the fixation cycle.

Zentbl. Bakt. ParasitKde. Abt. II, 138 : 154-156.

Four soil bacteria were grown in nutrient solutions containing N in the following forms : sodium nitrite, potassium nitrate, ammonium sulphate, azoxybenzene, hydroxylamine hydrochloride, alanine and aspartic acid, all at 2.5 and 5 mm. After 15 days, the N and C contents of the media were determined and the results interpreted in terms of N fixation.

97.
BHALLA, H. and SEN, A.N. (1971) Note on the stimulatory and inhibitory effects of some rhizosphere specific Rhizobium.

Indian J. agric. Sci. 41 : 1126-1127.

Certain rhizosphere isolates belonging to the genera Achromobacter Aerobacter, Agrobacterium, Bacillus, Micrococcus, Arthrobacter, Saricina and Pseudomonas had a stimulating effect; other isolates had an inhibitory effect on the Rhizobium of Bengal gram.

98.
BHANDARI, G.S.; DAHIYA, S.S. and SRIVASTAVA, O.P. (1971) The distribution and abundance of microflora in saline, saline-alkali, and alkali soils of Haryana State. Proc. All India Symp. Soil Salinity, Kanpur, 145-150.

Based upon seven soil profiles, salinity and alkalinity affected most adversely fungal flora, Calcareous soil with comparatively lower pH showed markedly higher count. Actinomycetes were not found to be significantly affected with moderate saline-alkali conditions and good organic carbon content; courts were higher than normal soil. Gross and bacterial populations in moderately saline-alkali soil with sandy loam texture of Rohtak were 64.1 and 68.9 per cent of normal soil respectively, but in other soils these were reduced. Nitrifying bacteria were affected more with the alkalinity and were found to be sharply decreasing with depth.

99.
BHARATI, S. G. and ANGADI, S.B. (1982) Long term effect of fungicides on nitrogen fixing blue-green algae. Indian J. Forestry 5 : 187-290.

The effects of Dithane Z-78 (Zineb) and Ceresan (Phenylmercury acetate) on two blue-green algae (cyanobacteria) are tabulated.

100.
BHARDWAJ, K.K.R. (1972) Note on the growth of Rhizobium strains of dhaincha (Sesbania cannabina (Retz.) Pers.) in a saline-alkaline soil. Indian J. agric. Sci. 42 : 432-433.

The survival and the proliferation of 4 efficient strains of Rhizobium of dhaincha isolated from 3 different soils were studied in a saline-alkali soil. The results indicated that the inoculant strains should be isolated from the problem soils in the interest of the establishment and proper multiplication of bacteria in the soil. Further certain amendments such as the application of gypsum and FYM may stand to ameliorate the adverse conditions to some extent for ecologically different rhizobial.

101.
BHARDWAJ, K.K.R (1972) Note on the occurrence of pigmented strains of Rhizobium in saline-alkali soil. Indian J. agric. Sci. 42 : 963-964.

Pigment-producing strains of Rhizobium of Sesbania bispinosa, Cyamopsis tetragonoloba, Melilotus indica and Trifolium alexandrinum were isolated from saline sodic soils.

102.
BHARDWAJ, K.K.R. (1973) Note on the occurrence of pigmented strains of Rhizobium in saline-alkali soil. Indian J. agric. Sci. 42 : 963-964.

The occurrence of relatively more pigmented rhizobia in salt affected soils is presumed to be an indication of some relationship between salt tolerance and pigmentation in these bacteria.

103.
BHARDWAJ, K.K.R. (1974) Growth and symbiotic effectiveness of indigenous Rhizobium species of salinealkali soil. Proc. Indian natn. Sci. Acad. 40B : 540-543.

Rhizobium species of Indian clover (Meltlotus parviflora Desf.), dhaincha (Sesbania aculeata (Retz.) Pers.) berseem (Trifolium alexandrinum juslen), guar (Cyamopsis tetragonoloba Taub.), cowpea (Vigna sienesis Savi.) and lentil (Lens esculenta Moench.) were found to be present in a highly saline-alkali soil. Good nodu- lation was obseved in Indain clover and dhaincha, poor in berseem, very poor in guar and cowpea and extremely poor in lentil. No nodulation was observerd in pea (Pisum sativum L.) Except those of Indian clover and dhaincha, majority of the rhizobial strains of other legumes were found to be symbiotically ineffective. Application of gypsum, farmyard manure and dhaincha hay to the soil had marked stimulatory effect on the growth and multiplication of the indigenous rhizobial strains.

104.

BHARDWAJ, K.K.R. (1974) Note on the distribution and effectiveness of rhizobium of Sesbania aculeata Poir. in saline-alkali soils. Indian J. agric. Sci., 44 : 683-684.

Nodulation and growth of Sesbania aculeata has been studied in 9 saline-alkali soils in Haryana, Punjab and Uttar Pradesh. It has been shown that the rhizobia of dhaincha are almost abundantly distributed in saline-alkali soils but their effectiveness varies widely in different soils and most of them appeared to be ineffective.

105.

BHARDWAJ, K.K.R. (1974) Numbers of bacteria in saline-alkali soils determined by a plate method. Soil Biol Biochem., 6 : 69-70.

A new medium for enumerating bacteria in salt affected soils has been formulated. The soil extract from a non-saline, non-alkaline soil was less effective than the extract of the saline-alkali soil in increasing the efficiency of the new medium, although extracts of each of the three salt affected soils viz. Mathura, Karnal and Indore, were equally effective.

106.

BHARDWAJ, K.K.R. (1975) Survival and symbiotic characteristics of rhizobium in saline-alkali soils. Pl. Soil. 43 : 377-385.

Survival, growth and symbiotic performance of rhizobia isolated from normal, saline-sodic and mildly acidic soils were studied in original and amended saline-alkali soils. Rhizobia of 4 out of 9 legumes studied for nodulation were found to be present in a highly saline-sodic soil. Majority of the strains of these bacteria did not survive in the original saline-sodic soil of pH 10.5 but as the pH was amended to lower than 10.0, all the strains survived in the soil. Virtually no differences were noticed in the survival and symbiotic characteristics of native and exotic strains of Rhizobium leguminosarum and Rhizobium trifolli in the saline-sodic soil, though wide variations were observed among individual strains irrespective of their ecological origin. Rhizobia were found to possess greater tolerance for alkalinity than their host-legumes. However, delayed nodulation in lentil (Lens esculenta) and berseem (Trifolium alexandrinum) resulting in decreased yield of the plants at pH values higher than 9.0, was observed.

107.

BHARDWAJ, K.K.R. (1978) Effect of soil pH and moisture level on growth, nodulation and nitrogen fixation by three legumes. pp.302. In S.P. Sen, Y.P. Abrol and S.K. Sinha (eds.), Nitrogen Assimilation and Crop Productivity. Associated Publishing Company, New Delhi.

Growth nodulation nitrogen fixation by dhaincha (Sesbania aculeata), sannhemp (Crotalaria juncea) and cowpea (Vigna sinensis) were studied under pot culture conditions at 8 soil pH values varying from 5.8 to 10.1 and 6 moisture levels ranging from 10 to 40 per cent.

Delayed nodulation was observed in all the three legumes at the highest pH limits restricting their growth. A relationship between proline accumulation in plant parts and delayed nodulation was noticed.

The growth and symbiotic characteristics of cowpea and sannhemp were affected more adversely at the higher than at the lower moisture levels. Nodulation and nitrogen fixation by dhaincha increased with the increase in soil moisture content upto 25 per cent and these characteristics remained unaffected upto the highest moisture level employed for the studies.

108.
BHARDWAJ, K.K.R. (1984) Process promoting cultures for composting.

Indian fmg. 33(5) : 17-19.

Results of the investigations on the use of nitrogen fixing and phosphate-solubilising cultures in composting convincingly revealed that treatment of the compost material with starter fungal culture of *Trichurus spiralis* or *Paecilomyces fusisporus* and the phosphate dissolving fungus *Aspergillus awamori* followed by inoculation with *Azotobacter chroococcum* in the presence of one per cent rock phosphate could yield better quality of compost than that obtained without these inputs.

109.
BHARDWAJ, K.K.R. and ABROL, I.P. (1978) Nitrogen management in alkali soils. pp. 83-96. In S.P. Sen, Y.P. Abrol and S.K. Sinha (eds.), Nitrogen Assimilation and Crop Productivity, Associated Publishing Company, New Delhi.

The ability of leguminous crops to fix atmospheric nitrogen is limited by the lack of effective strains of root nodule bacteria in these soils. Existing data on the efficiency of different sources of N, optimum quantity required for selected crops, transformations of applied fertilizers and extent of atmospheric nitrogen fixation in these problem soils is reviewed. Results of practical value are brought out and future research needs indicated.

110.
BHARDWAJ, K.K.R. and GAUR, A.C. (1968) Humic acid in relation to inoculation of crops.

Agrochimica 12 : 100-108.

In pot experiments involving inoculation with azotobacter or rhizobium and application of humates prepared from humic acid extracted from dung, application of 0.04% Na-humate to inoculated *Trifolium alexandrinum* markedly increased the yield and N content of the crop; Ca-humate applied to inoculated pea at 13.26 mg/litre nutrient solution had similar effects. High amounts of Ca-humate exerted a slight depressing effect on yield and N uptake.

111.
BHARDWAJ, K.K.R. and GAUR, A.C. (1971) Isolation and characterization of some humic acid decomposing bacteria and fungi from soil.

Zentbl. Bakt. ParasitKde. Abt. II, 126 : 307-312.

Organisms decomposing (decolourizing) Na humate in solid and liquid media were isolated, and identified as *Bacillus subtilis, B. brevis, Aspergillus fumigatus, A. niger* and *A. oryzae*, the fungi being more active than the bacteria.

112.
BHARDWAJ, K.K.R. and GAUR, A.C. (1971) Studies on the growth stimulating action of humic acid on bacteria.

Zentbl. Bakt. ParasitKde. Abt. II, 126 : 694-699.

Humic acid extracted from alluvial sandy loam was converted into Na-humate, and the effects of dialysed and autoclaved material (100-500 ppm) on *Azotobacter chroococcum* and *Rhizobium trifolii* were studied. The Na-humate did not serve as a source of Fe for *Azotobacter* but its effect on *Rhizobium* was comparable to that of yeast extract. Radiotracer studies with 14 C-labelled Na-humate indicated that there was no molecular absorption of the substance by cells of the test organisms.

113.
BHARDWAJ, S.P. and PATHAK, A.N. (1968) Effect of sulphur on bacterial nodulation of groundnut roots.

Curr. Sci. 37 : 351-352.

In pot experiments the number of nodules/plant at harvest increased from 9.3 to 121.3 as the S concentration in the nutrient solution was increased from 0-9 ppm and then decreased to 85.3 nodules/plant as the S concentration was increased to 81 ppm.

114.

BHARGAVA, J.K.; SAXENA, M.C. and TILAK, K.V.B.R. (1974) Response of soyabean to inoculation with various cultures of Rhizobium japonicum Proc. Indian natn. Sci. Acad. 40B: 471-475.

Field experiments carried out with different indigenous cultures of Rhizobium japonicm at Pantnagar revealed that inoculation with effective strain of R. japonicum was imperative for successful cultivation of soyabean. There was better nodulation and higher bacteroid and leghaemoglobin content in the nodules in the case of nodulation. Indigenous cultures E-188 (peat) E-188 (lignical), E-188 + E-400 (peat), SB-1 and SB-16 (peat) proved as effective as imported culture 'Nitragin'. Peat-based cultures were better than sand and broth cultures. Combination of effective strains improved the efficiency of inoculation as against single effective strain.

115.

BHARGAVA, S.S., RATHORE, K.S., SIAG, R.K. and MAKHAN LAL (1981) Response of Azitobacter under varying levels of nitrogen in bajra under unirrigated conditions.

Agric. Sci. Digest 1 : 133-134.

A field experiment was conducted at Agricultural Research Station, Durgapura during kharif 1976 to study the response of Azotobacter under varying levels of nitrogen in Bajra under unirrigated conditions. Seed inoculation with Azotobacter significantly increased the grain and straw yield. Increase in grain and straw yield due to seed inoculation with Azotobacter was 3.51 and 9.88 q/ha, respectively over no seed inoculation. Increasing levels of nitrogen also significantly increased the grain and straw yield. Maximum grain and straw yield 15.28 and 53.57 g/ha was recorded with the application 60 kg N/ha.

116.

BHARUCHA, F.R. and SHERIAR, K.C. (1952) An improved technique for estimating the nitrifying capacity of soils. Proc. Indian Acad. Sci. 35 B : 28-32.

Nitrifying capacity is best determined by measuring the NH_3, NO_2 and NO_3 contents of the soil at short intervals until the complete disappearance of NO_2. For aerobic soils the nitrifying capacity is inversely proportional to the time required for NO_2 disappearance.

117.

BHARUCHA, F.R.; DUBASH, P.J. and SHERIAR, K.C. (1951) Autocatalytic curve during nitrite formation.

Curr. Sci. 20 : 240-241.

When the values of nitrite accumulation in a fertile and an infertile soil were plotted against time an S-shaped curve was obtained which dropped sharply after achieving the peak, due to a rapid transformation of nitrite N to nitrate N after a certain threshold value. Oxidation of NH_4 in the fertile soil was faster than in the infertile soil.

118.

BHASKARAN, S. and VENKATARAMAN, G.S. (1958) Occurrence of blue-green alga in the nodules of Trifolium alexandrinum. Nature 181 : 277-278.

A blue-green alga, Nostoc punctiforme, occurred with rhizobial bacteria inside the nodules of T. alexandrinum. This occurrence in root nodules alone suggests that the relationship may not be simple endophytism; it is possible that alga may fix atmospheric N.

119.

BHASKARAN, T.R. and PILLAI, S.C. (1938) Production of the

mixed calcium salts of organic acids from cane molasses for fixation of nitrogen in the soil.
Proc. 25th Indian Sci. Cong. p 227.

Organic acids produced during the anaerobic fermentation of molasses induce N-fixation when added to soil. The fermentation product contains 10% of organic carbon. The N-fixed when the product is applied at the rate of 5 tons per acre corresponds to that in 200 kg of sulphate of ammonia.

120.
BHASKARAN, T.R. amd PILLAI, S.C. (1939) Nitrogen fixation in the soil by the solid lime compound prepared from cane molasses and its bearing on crop production. Proc. Soc. Biol. Chem. India 4 : 36-38.

Some results are given and the apparent transience of the N fixation is discussed. It is suggested that loss of biologically fixed N may occur generally in tropical soils.

121.
BHASKARAN, T.R. and SUBRAHMANYAN, V. (1937) Economy of carbon during decomposition of sugars in soil and its bearing on nitrogen fixation. Proc. 6th Ann. Conv. Sug. Tech. Assoc. India. p. 424-429.

Heavy inoculation with Azotobacter gave slow decomposition of sugar, there being little growth or N fixation when the sugar was exhausted. Mixed soil flora decomposed sugar rapidly, with little N fixation, but the organic acids formed served for growth and N fixation, the mechanism of fixation being different from that of Azotobacter.

122.
BHAT, J.V. and PALACIOS, G. (1948) Studies on the influence of some bacterial cultures in the nitrogen status of the soil. I - Influence of Serratia marescens in the nitrogen status of soils. II- Influence of Bacillus subtilis in the nitrogen status of the soil. J. Univ. Bombay 16B : 15-26.

Darkness and moisture were beneficial for N fixation by Serratia marescens. The organism did not survive for 30 days in the dry sample exposed to direct sunlight.

In a waterlogged sample Bacillus subtilis continuously changed insoluble N to soluble N, some of which was lost to the air. Losses in total N of the dry samples kept in darkness were smaller than those in samples kept in daylight, but conversion of total to soluble N was also much less, indicating decreased bacterial activity in darkness.

123.
BHAT, J.V. and PALACIOS, G. (1949) Influence of some bacterial cultures in the nitrogen status of soil. IV-Influence of Aerobacter aerogenes in the nitrogen stauts of soil. J. Univ. Bombay 17B : 84-87.
It appears that for Aerobacter aerogenes N fixation in the soil is favoured by moisture and absence of direct sunlight.

124.
BHAT, J.V. and PALACIOS, G. (1950) Influence of some bacterial cultures on the nitrogen status of soil. V-The influence of Staphylococcus citreus on the nitrogen status of soil. J. Univ. Bombay 18B: 17-20.

Dry semi-saturated and waterlogged soils were inoculated with measured amounts of young S. citreus culture suspensions and then kept in direct sunlight, diffused light or darkness for 30 days. The content of insoluble N increased in darkness, but decreased in the other samples, especially in the semi-saturated. Soluble N increased in all samples. Optimal conditions for a decrease in soil N were direct light and semi-saturation.

125.
BHATTACHARYYA, D.; MUKHERJEE, R. and MUKHOPADHYAY, A.K. (1982) Effect of insecticides on nodulation and rooting pattern of Phaseolus aureus L. and on microbial population in rhizosphere.

Trans. 12th int . Congr. Soil Sci. New Delhi 6 : 49.

Of the two insecticides applied to soil Lindane alone and together with FYM caused an increase in root volume and changes in the rooting pattern. Lindane alone significantly reduced both the size of the nodules and content of leghemoglobin. Thiodemeton did not have any such effect. The bacterial population is significantly reduced initially (upto 10 days) upon treatment with both the insecticides alone and in combination with FYM. After 10 days the bacterial population slightly increased, but not significantly, except in the case of Thiodemeton plus FYM, which showed a significant increase. No significant effect on the nitrogen and phosphorus status of soil was observed as a result of treatment with the two insecticides.

126.
BHATTACHARYA, R. (1958) Growth of Azotobacter in rice soils. Indian J. agric. Sci. 28 : 73-80.

More than 50% of the samples of rice soils collected from different parts of West Bengal contained no Azotobacter, thought the majority of soils had favourable pH values of @ 6. The numbers of Azotobacter in the remaining soils were small, and a rapid disappearance of the organism occurred when added to some rice soils. There was no evidence of antibiotic or toxic substances in the soils, but the growth of Azotobacter appeared to be depressed, especially in the subsoil by waterlogged conditions.

127.
BHRIGUVANSHI, S.R. and GANGWAR, B.R. (1984) Effect of temperature on the survival of Rhizobium species.

J. Indian Soc. Soil Sci. 32 : 172-174.

A laboratory experiment was carried out with FYM, leaf mould and berseem grain meal as carrier and two types of soil having different texture. FYM helped in better survival condition of the organism when exposed to super-optimal temperature (@ 40ºC) and berseem grain meal seemed to have an edge over it. R. trifolii grew better in the presence of leaf mould mixed with heavy soil.

128.
BHURAT, M.C. and SEN, ABHISWAR (1968) Nitrogen-fixing bacteria in the phyllosphere of wheat and pea. Indian J. agric. Sci. 38 : 319-325.

Leaf surfaces of wheat and pea were observed to harbour nitrogen-fixing organisms of the genera *Achromobacter*, *Pseudomonas* and *Cellulomonas*. Leaves of pea were found to have, in addition, a *Bacillus* species. Nitrogen fixation by the microorganisms occurring in the leaves of wheat varied from 7.6 to 13.6 mg and that by the micro-organisms in the leaves of pea from 6.9 to 15.0 mg per gram of sugar in 14 days. Spraying the crops with 1 per cent sugar solution every alternate day increased their yields significantly.

129.
BHURAT, M.C. and SEN, A. (1971) Comparison of microbiological activities of the phyllosphere and rhizsphere of some agricultural crops at the preflowering stage.

Zentbl. Bakt. ParasitKde Abt. II, 126 : 8-13.

Nitrification in the phyllosphere was restricted to summer crops only. Denitrification was observed only in maize, paddy and cluster bean and varied from 1.16 (wheat) to 6.53 (barley) mg/g sugar. The phyllosphere of winter legumes, pea and gram solubilized much less phosphate than summer legumes (guar, groundnut). Ammonification by the rhizosphere varied from 0.37 (wheat) to 0.75 (paddy) g NH_4-N/g peptone-N. There was no nitrite formation by the rhizosphere of summer crops; denitrification activity was absent in rhizosphere soils of winter crops. N fixation, P solubilization and cellulose decomposiition were higher in the rhizosphere than in the phyllosphere.

130.
BHUVANESWARI, K. and SUBBA RAO, N.S. (1957) Root exudate in relation to the rhizosphere effect. Proc. Indians Acad. Sci. 45 : 299-301.

Tartaric and oxalic acids, d-xylose and d-fructose were identified in exudates of _Sorghum vulgare_ var. _dochna_ and _Brassica juncea_ (mustard). Mustard, which shows a reduced rhizosphere microflora, had also malic and citric acids, d-glucose and maltose.

131.
BISHNOI, K.C. and RAM DUTT (1983) Studies on the effect of methods of inoculation on yield and yield attributes under varying nitrogen levels.

Indian J. Agron. 28 : 79-81

Results show that the differences in the grain yield between the strains was not significant. Different methods of inoculation markedly influenced the seed yield. Both soil as well as seed inoculation gave significant yield increase over no inoculation.

132.
BORTHAKUR, H.P. and SARMAH, S.C (1983) Effect of _Rhizobium_ and _Azospirillum_ cultures on the growth and yield of soyabean.

Seeds & Farm 9 (6) : 30-31.

Results show that application of nitrogen in the form of urea and seed inoculation with _Azospirillum_ culture stimulated the native strain of _Rhizobium_ for nodule formation in soyabean. Even the heat killed _Rhizobium_ and _Azospirillum_ cultures could exert a stimulatory effect of considerable magnitude on the native _Rhizobium_ strain for nodule formation in the root system of soyabean.

133.
BOSE, S.R. (1943) Fixation of atmospheric nitrogen by fungi. _Sci. Cult._ _8 : 389._

Statements of previous authors that mycorrhizal fungi can fix atmospheric nitrogen do not stand up under experimentation. The only fungi able to fix gaseous N from the atmosphere are probably a few species of _Phoma._

134.
CHAHAL, D.S. and GILL, S.S. (1974) Studies on oxygen uptake of cultured cells and bacteriods of Rhizobium sp. from mung (Phaseolus aureus Robx.). Proc. Indian natn. Sci. Acad. 40B : 599-603.

The highest oxygen uptake by cultured cells is observed with succinate (48.8 $\mu l O_2$/mg dry wt cells/hr) followed by glucose, pyruvate, ribose, α ketoglutarate and lowest with oxalte (6.2 $\mu l O_2$/mg dry wet cells/hr) Similar trend was observed with bacteroids. In general, the oxygen uptake was higher in case of cultured cells than that in the bacterioids on the substrates studied.

Oxygen uptake in the presence of the inhibitors at the concentrations of 0.001 to 0.1M was pH dependent. Some stimulation in oxygen uptake by both cell types at lower concentration of sodium fluoride and at all concentrations of malonate but only for cultured cells, was observed.

135.
CHAHAL, V.P.S. and SHARMA., P.K. (1982) Use of lentil nodule on an antigen for strain identification of Rhizobium leguminosarum.

Zentb. Mikrobiol. 137 : 118-120.

Lentil seeds were inoculated with the rhizobia, and crushed nodules from the subsequent plants were used as the antigen in agglutination tests. A pot culture experiment was conducted to observe the establishment of eight efficient strains of Rhizobium leguminosarum in competition with native rhizobia in unsterilized soil. The strains varied considerably in their ability to compete with native rhizobia to form nodules. Good competitor strains PL 29 and PL 36 of pea resulted in poor nodulation of lentil. Strain PL 39 proved the best competitor as it formed 76.4% of the total number of nodules. Four serogroups were recognizable among the strains.

136.
CHAKRABORTY, T. and SEN, S. (1967) Mineralization of carbon from added organic matter in the soil under laboratory conditions.

J. Indian Soc. Soil Sci. 15 : 153-155

Under high temperature and favourable moisture conditions, the decomposition of green matter in the soil is rapid and three is very little gain of organic carbon, except the maintenance of reserve carbon level in the soil.

137.
CHANDRA, A.K. (1974) Induction of mutation in Rhizobium japonicum by Ultraviolet Radiation and selection of agronomically superior serotypes. Proc. Indian natn. Sci. Acad. 40B : 491-498.

Fifteen serotypes of Rhizobium japonicum. BU/S4, were obtained by ultraviolet radiation and compared with the wild strains for their biochemical properties and biological efficiency of nitrogen fixation. Indications for improving specific characteristics of soyabean plants by specific serotypes have been obtained.

138.
CHANDRA, S.; RAIZADA, M. and KHANNA, K.K. (1981) Effect of organic soil amendments on the rhizosphere microflora of tomato.
Proc. Indian Acad Sci. 90 : 189-197.

In pot experiments, the effect of various organic amendments (dry straws, young fresh leaves, whole fresh plants) on the microbial population of rhizosphere and non-rhizosphere soils was studied. The amendments had marked stimulatory or inhibitory effects on the rhizobial population of fungi and bacteria. The effect varied with the amendment and with the variety and age of the plant. The materials also affected the relative population of different fungi in the rhizosphere. A number of them suppressed the population of Fusaria and stimulated that of Aspergilli.

139.
CHANDRAMOHAN, D. and MAHADEVAN, A. (1968) Indole acetic acid metabolism in soils.

Curr. Sci. 37 : 112-113.

Soil differed in their ability to convert tryptophane to IAA during incubation with sucrose and a phosphate buffer. Maximum and minimum amounts of IAA were synthesized by a clay soil and a river sand, respectively. Addition of streptomycin had little effect on IAA-synthesizing ability. The soils had greater IAA-synthesizing activity than IAA destroying ability, but it is believed that the IAA synthesized by soil micro-organisms contributes little to the IAA pool of the plant.

140.
CHANDRASEKHARAN NAMBIAR, P.T. and SHETHNA, Y.I. (1977) Effect of NH_4 + on acetylene reduction (nitrogenase) in *Azotobacter vinelandii* and *Bacillus polymyxa.*

J. Indian Inst. Sci. 59 : 155-158.

In culture medium, NH_4^+ inhibited the nitrogenase activity of *A. vinelandii* immediately, but did not affect that of *B. polymyxa*. Inhibition in *A vinelandii* was reversible in the early stages of incubation with NH_4^+ but irreversible during later stages.

141.
CHANDRASEKARAN, S. (1969) Production of organic acids by soil micro-organisms.

Pl. Soil 30 : 299-304.

Micro-organism from the rhizospheres of different plants (sesbania, cowpea, sorghum, millet) and from non-rhizosphere soil were incubated in glucose-containing liquid media for a fortnight. Chromatographic analyses showed that rhizosphere organisms produced greater amounts of acids (tartaric, citric, succinic, lactic) than organisms from non-rhizosphere soil did greater amounts of acids were produced by rhizosphere organisms from legumes than from non-legumes. There were no qualitative differences in the acids produced by micro-organisms from the rhizosphere, and from non-rhizosphere soil.

142.
CHARYULU, P.B.B.N. and RAJARAMAMOHAN RAO (1980) Influence of various soil factors on nitrogen fixation by Azospirillum spp. Soil Biol. Biochem. 12 : 343-346.

The influence of soil pH, redox potential and added organic matter on N_2-fixation by Azospirillum was studied. Application of rice straw to submerged alluvial, laterite and acid-sulphate Pokkali soils enhanced the population of N_2-fixing Azospirillum spp. An acid-sulphate saline soil (pH 3.2) harboured Azospirillum spp with appreciable N_2-fixing activity. Enrichment cultures originating from soils with low pH (< 4.0) possessed lower N_2-fixing activity than cultures from soils with higher pH values (up to 6.6). Azospirillum cultures from soils that had undergone prolonged waterlogging showed lower N_2-fixing activity than cultures isolated from soils submerged for a few days. A relationship was shown between the in vitro N-fixing activity of Azospirillum cultures and the redox status of the soil samples; activity was high when the soil redox potential was between-50-150mV.

143.
CHARYULU, P.B.B.N. and RAO, V.R. (1981) Influence of ammonium nitrogen on nitrogen fixation in paddy soils. Soil Sci. 131 : 140-144.

The effect of ammonium sulfate on heterotrophic nitrogen fixation in four paddy soils under two water treatments was studied employing acetylene reduction assay. In alluvial, laterite, and acid saline (Karapadam) soils amended with cellulose, more nitrogen fixation occurred under nonflooded conditions (60 per cent water-holding capacity) than under flooded conditions. In submerged soils, nitrogen fixation decreased with increasing concentration of ammonium sulfate; under nonflooded conditions, this inhibition was more pronounced. The alteration in the acetylene reduction activities were attributed to the differential degree of organic matter decomposition under the two water regimes. The maximum acetylene reduction was noticed during the 9-12 day incubation irrespective of water regime. Low concentrations (5 to 20 ppm) of added N stimulated nitrogen fixation, though higher concentrations were inhibitory. Nitrogen fixation in acid soils was lower than in alluvial and laterite soils.

144.
CHARYULU, P.B.B.N. and RAO, V.R. (1981) Influence of carbon substrates and moisture regime on nitrogen fixation in paddy soils. Soil Biol. Biochem. 13 : 39-42.

Heterotrophic N_2 fixation in four paddy soils under flooded and non-flooded conditions was greater when soils were amended with cellulose and rice straw, the former being superior. Addition of sucrose and glucose stimulated N_2 fixation in alluvial soil and non-flooded acid saline soil, while sucrose alone enhanced N_2 fixation in laterite soil. Malate stimulated N_2 fixation in flooded alluvial soil only. The populations of nitrogen fixing micro-organisms varied considerably among soil types.

145.
CHARYULU, P.B.B.N.; NAYAK, D.N. and RAO, V.R. (1981) 15N

2 incorporation by rhizosphere soil. Influence of rice variety, organic matter and combined nitrogen.

Pl. Soil 59 : 399-405

Heterotrophic nitrogen fixation by rhizosphere soil from 20 rice cultivars was estimated employing the 15N-tracer technique. Rhizosphere soil samples from different rice cultivars showed striking difference in $15N_2$ incorporation, the amount of nitrogen fixed ranging from 1.7 to 16.5 mg N_2/kg soil. Rhizosphere samples from soil amended with rice straw (3 and 6 tons/ha) exhibited more pronounced nitrogen fixing activity than the samples from unamended soil, while the activity of the rhizosphere samples was suppressed by combined nitrogen (40 and 80 kg N/ha). The inhibitory effect of combined nitrogen was not expressed in the presence of 6 tons/ha rice straw.

146.
CHATTERJEE, M. and CHATTERJEE, S.P. (1983) Nitrogen fixing cyanobacteria from the paddy field soils of Burdwan district. Phykos 22: 64-66.

Twenty eight soil samples from rice field of 14 different places of Burdwan district (W.B.) were tested for their N-fixing cyanobacteria. 133 isolates belonging to 21 species under 7 genera of blue-green algae were obtained. Species of Anabaena and Nostoc had widespread occurrence.

147.
CHATTERJEE, S.K. and NANDI, B. (1981) Biodegradation of wheat stubbles by soil micro-organisms and role of the products on soil fertility.

Pl. Soil 59 : 381-390.

Soil fungi and actinomycetes degraded wheat stubble under laboratory conditions. Mixtures of species were more efficient than individual species in degrading lignin, holocellulose and nitrogenous substances. Degradation of lignin and holocellulose caused an increase in the content of soluble carbohydrates, while degradation of nitrogenous substances caused an increase in the soluble nitrogen content of the degraded product. The residues of stubble degradation improved the growth of maize and rice.

148.
CHAUDHURI, S. and MATTI, S. (1978) Inhibitory activity of inorganic nitrogen sources to sclerotia of Sclerotium rolfsii.

Z. Pflan. Pflanzenschutz 85 : 10-14.

Nitrite and ammonium at > 2000 ppm caused irreversible loss of viability of S. rolfsii sclerotia in vitro, whereas nitrate at these or higher concentrations caused reversible toxicity. Fungal growth from active mycelial inoculum was not affected by the N sources at rates toxic to sclerotia.

149.
CHAUHAN, S.K.; TYAGI, V.K. and NAGAR, M.L (1981) Mycoflora of soil around pneumatophores of Soneratia acida L. in Andaman Islands.

J. Indian Bot. Soc. 59 : 281-285.

Of the 60 fungal species recorded in soil around pneumatophores of the mangrove Soneratia acida, 21 were common at all three sites, the most prevalent genera being Aspergullus, Penicillium, Fusarium, Trichoderma, and Rhizoctonia.

150.
CHENDRAYAN, K,: ADHYA, T.K. and SETHUNATHAN, N. (1980) Dehydrogenase and invertase activities of flooded soils. Soil Biol. Biochem. 12 : 271-273.

The dehydrogenase and invertase activities of three soils were studied under flooded and non-flooded (60% water holding capacity) conditions. Flooding increased the dehydrogenase activity, but decreased the invertase activity. Addition of rice straw increased the invertase activity only under flooded conditions.

151.
CHHONKAR, P.K.; SUBBARAO, N.S. (1966) Fungi associated with legume root nodules and their effect on rhizobia.

Can. J. Microbiol. 12 : 1253-1261.

Fungi associated with the nodules of nine common legumes were isolated and identified. Nodule mycoflora comprised Cephalosporium, Alternaria, Aspergillus, Penicillium, Rhizopus, Acrothecium, Fusarium, Rhizoctonia, Curvularia, Pythium and Trichoderma species. Composition of the mycoflora of Trifolium alexandrinum nodules varied with age of plant; it also varied a little with volume of nodules. The fungi were screened for antibiotic activity towards six species of Rhizobium; R. trifolii,

R. phaseoli, R. leguminosarum and R. japonicum were more susceptible than R. meliloti or Rhizobium sp. (of the cowpea group). Myco-colonization in soil by an isolate of Cephalosporium sp. significantly reduced the fresh weight and N status of T. alexandrinum.

152.
CHHONKAR, P.K. and SUBBA RAO, N.S. (1967) Phosphate solubilization by fungi associated with legume root nodules.

Can. J. Microbiol. 13 : 749-753.

Phosphate-solubilizing ability of different isolates of fungi associated with legume root nodules was studied vitro. Among the fungi tested, isolates of Penicillium lilacinum, Aspergillus sp,. A flavus, A. niger, A. terreus and A. nidulans solubilized $Ca_3(PO_4)_2$. When soluble KH_2PO_4 was present with Ca_3-$(PO_4)_2$ in the medium, some of the fungi failed to solubilize phosphate.

153.
CHOPRA, R.K. (1983) Effect of water stress on photosynthesis and nitrogen fixation in legumes. Proc. Natl. Symp. Adv. Front. Pl. Sci. Jodhpur 7 : 79-80.

Legumes are characterized by their ability to fix atmospheric nitrogen in symbiosis with Rhizobium. Water stress reduces N-fixation by directly affecting nodule physiology and indirectly by reducing the availability of photosynthates.

154.
CHOUDHURI, D.K. and ROY, S. (1967) Qualitative composition of the collembolan fauna of some uncultivated fields in Nadia district, (West Bengal) with a correlation between monthly population and individual soil factor.

Rev. Ecol. Biol. Sol. 4 : 507-515.

Samples taken from nine uncultivated fields showed that maximum collembolan populations occurred in July, August and

September and minimum populations in February and April. Soil factors (pH, water content, organic C, $CaCO_3$, phosphate and nitrate content, and particle size) of the Ganga alluvial soil were studied. Organic-C content and soil moisture varied with the season and their effect on population fluctuation was significanty; particle size also showed some positive correlation. Populations were highest in the upper layers and declined with depth.

155.
CHOUDHURI, D.K. and ROY, S. (1968) Qualitative composition of the Collembolan fauna of some uncultivated fields in West Bengal with correlation between monthly population and different soil factors. In Progress in Soil Biodynamics and Soil Productivity (A. Primavesi, ed.), Pallotti, Santa Maria, Brazil, 176-177.

The Collembola population was maximal during the monsoon months. Dominant genera and families are listed. Organic C, moisture and P content were positively correlated with Collembola population.

156.
CHOWDHURY, S. (1946) A brief review of recent observations regarding soil micro-organisms. Allahabad Farmer 20 : 19-28.

Methods are described of the use of micro-organisms as test plants to estimate soil fertility and detect deficiency in soil nutrients such as P and K. The principal factor responsible for variation in the numbers of micro-organisms in different soils and at different depths, is the nature and amount of organic matter. Soil reaction is of secondary importance.

157.
DADARWAL, K.R. and SEN, A.N. (1971) Survival of Rhizobium japonicum on soyabean seeds.

Indian J. agric. Sci. 41 : 564-568.

Rhizobium japonicum applied in the form of liquid culture to unsoaked soyabean seeds decreased rapidly until only 2% survived after 2 days. If the seeds received a preliminary soaking in sterile water overnight, or peat-based cultures were used for inoculation, Rhizobium increased greatly in numbers. Preliminary soaking removed the toxic seed diffusates increasing the number of Rhizobium cells by 200-600% after 2 days. After 7 days the viable count had decreased in all cases, but 30-70% still remained on soaked seeds inoculated with peat-based culture.

158.
DADARWAL, K.R. and SEN, A.N. (1974) Varietal specificity for Rhizobial serotypes in relation to nodulation and crop yield. Proc. Indian natn. Sci. Acad. 40B: 548-553.

The nodulation patterns and yield with native and introduced rhizobia (as distinguished serologically) were examined in different genotypes of bengalgram (Cicer arietinum L.) pea (Pisum sativum L.) soybean (Glycine max L. Merr.), green-gram (Phaseolus aureus Roxb.) and some of the radiation induced mutants of green-gram in M_3 generation, under natural green house conditions. It was found that under similar soil and environmental conditions with similar doses of inoculum, the number of nodules produced and the efficiency of an inoculating strains varied with the genotype of the host species. Both nodulation and response to yield were found to be quantitative characters. Possibility of improving those characters through mutation breeding is discussed. The time of flowering and maturity of the host were not related with the efficiency.

The nodule bacteriods showed homology for somatic antigen with cultured rhizobia but differed to internal protein antigen bands.

159.
DADARWAL, K.R.; SHASHI PRABHA and TAURO, P. (1978) Varietal differences with regard to Rhizobium compatibility and efficiency of nitrogen fixation in chick pea (Cicer arietinum) pp. 235-243. In S.P. Sen, Y.P. Abrol and S.K. Sinha (eds.), Nitrogen Assimilation and Crop Productivity. Associated Publishing Company, New Delhi.

A detailed study was carried out to determine differences in antigenic properties, antibiotic sensitivity, host compatibility and effeciency among native rhizobia associated with different varieties of chick pea. It was found that rhizobia isolated from different varieties grown at one locality were antigenically distinct, differed in their ability to nodulate and fix nitrogen. It was also found that nodulation in chick pea depends to a great extent on the host genotype. A positive correlation between growth rate and resistance to penicillin and streptomycin in this group of rhizobia was observed.

160.
DHADHICH, K.S; VARMA, A.K. and VENKATARAMAN, C.S. (1969)
The effect of Calothrix inoculation on vegetable crops.

Pl. Soil 31 : 377-379.

Inoculation of the soil with Calothrix anomale had a positive effect on Capsicum annuum (Chillies) and Lactuca sativa, in terms of yield and nitrogen content. A combined application of urea and blue-green alga was more effective than individual applications.

161.
DAHIYA, S.S.; BHANDARI, G.S. and SRIVASTAVA, O.P. (1976)
Distribution and abundance of microflora in salt affected soils of Haryana State.

J. Res. HAU, 6 : 6-11.

Seven soil profiles representing saline, saline-alkali, alkali and normal soils were analysed for physico-chemical characteristics and distribution and abudance of microorganisms. The different groups according to dominance in these soils were : total bacteria, nitrosomonas bacteria, actinomycetes and fungi. Fungal flora were the maximum sufferer of salinity and alkalinity, however, calcareous soil of Hathwala with comparatively lower pH, showed markedly higher count. Actinomycetes were not found to be significantly affected with salinity and alkalinity, rather in some soils of moderately saline-alkali condition and good organic corbon cantent i.e., Rajthal and Rohtak, counts were higher than normal soils. Bacterial population in a moderately saline-alkali soil, with sandy loam texture of Rohtak were 64.1 and 68.9 per cent of normal soil, respectively, but in other soils these were highly reduced, Nitrosomonas bacteria were more affected with the alkalinity (Karnal and Bhambhewa soils) and were found to be sharply decreasing gradually with the depth while actinomycetes and fungi were almost uniformly distributed and did not show a regular trend.

162.
DAJI, J.A.; PATEL, M.K. and KIBE, M.M. (1951) Molasses and mohua flowers as sources of energy for Azotobacter. Proc. Indian Acad. Sci. 33B : 322-326.

Both the materials supplied energy to the Azotobacter, but when the amount of material added to Ashby solution used as culture medium exceeded 1 g, the amount of N-fixed decreased and gradually ceased. The presence of available N in the medium depresses the N-fixing power Azotobacter and it is possible that it may also be inhibited by the decomposition products of the sugar which are mainly acidic.

163.
DANDEKAR, A.M.; VASAVADA, H.A. and MODI, V.V. (1978) Partial purification of a competence factor from Rhizobium japonicum.

Arch. Microbiol. 118 : 253-256.

A competence factor (CF) from Rhizobium japonicum was partially purified to 43 fold on Sephadex G-100. This CF preparation was sensitive to heat, trypsin and pronase and was resistant to DNase 1, RNase A and lysozyme. It had an approximate mol. wt. of 82,000. Osmotic shock treatment of competent cells revealed that the CF is located in the periplasmic region of the cell.

164.
DART, P.J. (1978) Nitrogen fixation in the semi-arid tropics.

pp.288. In S.P. Sen, Y.P. Abrol and S.K. Sinha (eds.), Nitrogen Assimilation and Crop Productivity. Associated Publishing Company, New Delhi.

Little fertiliser is applied in the semi arid tropics, and hence nitrogen inputs come mainly from biological nitrogen fixation. Legumes such as chickpea, pigeon pea and groundnut fix some nitrogen but little is known of the benefits of this for subsequent crops. Cereals such as millet, sorghum and rice stimulate nitrogen fixation by bacteria associated with their roots. Blue green algae also contribute some fixed nitrogen.

165.
DAS. A.C. (1963) Ecology of soil fungi of rice fields. I- Succession of fungi on rice roots. II - Association of soil fungi with organic matter. Trans. brit. Mycol. Soc. 46: 431-443.

Species of *Aspergillus* and *Pennicillium* are common in the early stages of rice growth, but specialized root inhabitants such as *Thielavia angulata* and *Humicola* sp. are dominant on mature roots; *Trichoderma* and *Cephalosporium* sp. tend to become prominent on dead and decomposing roots after harvest. The greater frequency of isolation of some fungi from washed organic matter is due to their greater saprophytic ability to colonize organic debris in the soil.

166.
DAS, A.C. (1963) Utilisation of insoluble phosphates by soil fungi.

J. Indian Soc. Soil Sci. 11 : 203-207.

Eighteen fungi isolates from soil of a paddy field had been tested for their ability to utilise insoluble inorganic phosphates in cultures. In agar media comparable growth were notes for all, both in soluble and insoluble phosphates, but clear zones around the colonies were found only in the cases of *A. niger*, *S. rolfsii* and *Penicillium*-10. By comparing the dry weights in liquid cultures, nine fungi were noted to be strong utilisers of all the insoluble phosphates supplied with; four were found to exhibit preferences to some insoluble froms, while three of them seemed to be weak utilisers. The remaining two T viride and Syncephalastrum sp. did not produce appreciable amount of dry matter in any of the media employed. Acidity of culture media could not be correlated with phosphate utilisation.

167.
DAS, A.C. (1966) A new fungus from rice field soils.

Technology (India), 2 : 245-246.

Humicola oryzae sp. nov. is described. It was isolated many times from paddy roots, but rarely from soil.

168.
DAS, A.C. and KHAN, J.A. (1967) Ureolytic bacteria in soils. Part III. Volatilization loss of ammonia from urea on soil application.

Technology (India), 4 : 16-18.

When urea was applied uniformly to the surface of paddy soil brought to suitable granule size and moisture content, volatilization losses of 25.2-30.91% at 30°C and 3.4-10.0% at 18-25°C during the subsequent 15-day period were noted. If urea was applied a few inches below the surface, losses were negligible. Though water-logging and acidic soils tend to minimize the loss to some extent, soil-moisture contents of 40 and 60% saturation had no such effect.

169.
DAS, S. N. and BADURI, P.N. (1974) Host-Rhizobium interaction between Phaseolus vulgaris L. and Rhizobium Phaseoli. Proc. Indian natn. Sci. Acad. 40B : 554.561.

During the study made on inter-strain competition for nodule sites between four serodistinct strains of R. phaseoli were taken at a time in different combinations as inoculum. The cell population of strains in the mixtures were kept equal in preposition. The host varieties were inoculated simultaneously by the individual strains as control.

Presence of a new strains may affect the competitiveness of one strain over the other. Differential preferentialities towards different strains of R. phaseoli constitute a primary limiting factor in inter-strain competition for nodule sites in different varieties of P. vulgaris.

170.
DAS, S.K.; BALASUBRAMANIAN, V. and RAO, A.C.S. (1984) Azospirillum in nitrogen economy of rainfed sorghum and pearl millet. Golden Jubilee Seminar on Soil Resources and Productivity Management, Indian Society of Soil Science, New Delhi 3 : 56-57.

In order to find out alternate sources of nitrogen in dryland farming an experiment was conducted for two years 1981-82 during kharif (monsoon) season on a semi-arid red sandy loam soil of Hyderabad to test the efficiency of Azospirillum brasilence on sorghum and pearl millet with and without fertilizer nitrogen. During 1981, on as average, 18.0 and 23.0 per cent increase in grain yield was obtained due to seed inoculation with Azospirillum for sorghum and pearl millet, respectively. The increase in yield due to inoculation was found to be equivalent to 20 kg fertilizer nitrogen per hectare. However, the response of Azospirillum inoculation in kharif 1982 was of the order of 7.4 per cent for sorghum and 10.4 per cent for pearl millet. This low response in the second year might have been due to unfavourable rainfall (190 mm during the cropping seasons for both sorghum and pearl millet) as against much favourable rains during 1981 (730 mm for sorghum and 590 mm for pearl millet). The interaction between fertilizer nitrogen and Azospirillum inoculation was non-significant in both the years. Data collected in 1981 on leaf development, growth and number of grains per cob showed significant beneficial effect of Azospirillum inoculation in both the crops. The nitrogen content in plants was not found to be affected by inoculation, while the total nitrogen uptake increased by nitrogen application with or without inoculation.

171.
DASGUPTA, J.; DINKAR, A.W. and KANNAN, L.V. (1970) This distribution of antagonistic actinomycetes of three types of Indian soils.

Folia microbiol. 15 : 368-371.

Antibiotic properties of actinomycetes (including antibiotic activities and antibiotics produced) isolated from hill, alluvial and lateritic soils are discussed and summarized in tables.

172.
DASH, M.C. and THAMBI, A.V. (1978) A taxonomic study of Enchytracidae (oligochaeta) from grassland soils of southern Orissa, India.

Rev. Ecol. Biol. Sol. 15 : 129-134.

The following species of Enchytraeidae are described from tropical grassland and arable soils of southern Orissa, India : Marionina indica, M. macfadyeni Hemienchytracus Khallikotosus, Enchytraeus berhampurosus, Achaeta nurmineni. All are new species. The genus Marionina is reported for the first time in India.

173.
DASH, M.C.; MISHRA, P.C.; MOHANTY, R.K. and BHATT, N. (1981) Effects of specific conductance and temperature on urease activity in some Indian soils. Soils Biol. Biochem. 13 : 73-74.

Urease activity in pasture, forest and hill soils at Orissa, India, was positively correlated with total N content, organic C content and specific conductance, indicating that organic matter content accounted for most of the variation in urease activity. Activity was also positively correlated with silt + clay content, and negatively correlated with moisture content, pH and sand content. Maximum urease activity occurred at a soil temperature of 47°C; in these soils the temperature at 3 cm depth reaches 45°C during May (summer). Urease activity, specific conductance and organic C and total N contents decreased with increasing soil depth.

174.
DASH, M.C.; NANDA, B. and BEHERA, N. (1980) Fungal feeding by Enchytraeidae (Oligochaeta) in a tropical woodland in Orissa, India.

Oikos 34 : 202-205

The food preference of three species of Enchytraeidae were examined by three techniques : (1) squash preparations; (2) culturing of gut contents; (3) selective feeding on pure fungal cultures. Enchytraeids ingested large amount of organic material and fungi. Fungal material formed some 30% of the gut contents. Anteclitellar part of the body contained significantly higher per cent of plant and fungal material than the postclitellar part. Analysis of gut contents showed that Rhizopus nigricans, Syncephalastrum racemosum, and Trichoderma viride are generally digested by enchytraeids while Penicillium steckii and Aspergillus niger are not digested. Isolation made from postclitellar gut content did not generally give rise to fungal colonies except for Aspergillus niger and Penicillium steckii.

175.
DATTA, T. (1979) Effect of organic and inorganic compounds and carbon dioxide in the excystment of soil amoebae.

Arch. Protistenkunde 121 : 155-161.

In the laboratory, acetates and nitrates of sodium, potassium and copper induced excystment of Schizopyrenus russelli and Hartmanella rhysodes cysts to varying degrees. A minimum atmospheric concentration of CO_2 of 0.03% was essential for excystment. Slight increase in the CO_2 level (0.5-0.8%) enhanced excystment, but a further increase in CO_2 inhibited the excystment of amoebae. Various concentrations of sugars and antibiotics were unable to induce excystment.

176.
DATTA, T., GHAI, I.; SOOD, M.L. and GAUTOM, R.K. (1982) Effect of fertilizers and farmyard manure on protozoa population of field soil. J. Indian Soc. Soil Sci. 30 : 146-149.

The total number (active + cystic) and the number of active protozoon in plots treated with NPK alone and in various combinations and NPK + FYM were much higher than in untreated plots. Total accounts of amochae in mineral treated plots did not differ significantly from those in the FYM plots, except the plot where $ZnSO_4$ was applied alone. Positive correlation was found between population of amoebae and the crop yield.

177.
DAYAL, R. and GIRI, L. (1971) Soil fungi in leguminous fields of Varanasi, India.

<u>Oikos</u> 22 : 122-127.

The distribution of 67 species of fungi (9 Phycomycetes, 12 Ascomycetes, 45 Deuteromycetes and 1 · Basidiomycete) under <u>Crotalaria juncea, Cicer arietinum</u> and <u>Cajanus cajan</u> was studied in relation to moisture, pH, depth of sampling and season.

178.
DAYAL, R. and GUPTA, O.S.D. (1968) The soil fungi of Varanasi, India, in relation to the surface vegetation.

<u>Oikos</u> 19 : 139-142.

76 , 70 and 60 species of fungi were isolated from soil under fallow, soil from the bank of a kachcha pond and bare soil in a drainage channel, respectively. The Deuteromycetes and Phycomycetes Yielded the greatest numbers of species at all three sites.

179.
DAYAL, R. and SRIVASTAVA, S.S (1979) Three catenate conidial phycomycetes destructive to soil amoebae.

<u>Indian Phytopath.</u> 32 : 397-400.

Three species of Zoopagales forming Catenate conidia belonging to the families Zoopagaceae and Cochlonemaceae, hitherto unrecorded from India, were found killing and consuming amoebae from Varanasi soil. The morphology of Zoopage tryphera Drechsler, Zoopage thamnospira Drechsler and Cochlonema megalosomum Drechsler is described and illustrated.

180.
DE, P.K. (1939) The role of blue-green algae in nitrogen fixation in rice fields. Proc. Roy. Soc. Ser. B. 127 : 121-139.

Evidence showed that algae are the main agents of nitrogen fixation in rice fields and that bacteria are not important.

181.
DE, P.K. and MANDAL, L.N. (1965) Fixation of nitrogen by algae in rice soils. Soil Sci. 81 : 453-458.

In cropped unfertilized soils N fixation varied from 13.8 to 44.4 kg/ha 60 kg/ha P_2O_5 stimulated fixation only in soils poor in P. In some soils there was evidence that the application of Mo + P caused greater N fixation than that of P alone. More than 80% of the N-fixation occurred in the first 4 weeks after transplanting.

182.
DE, P.K. and SULAIMAN, M. (1950) Fixation of nitrogen in rice soils by algae as influenced by crop, CO_2 and inorganic substances. Soil Sci. 70 : 137-151.

Increased development and N-fixation by algae occurred in the presence of the crop and was enhanced by N-free inorganic fertilizers. In soils of pH less than 6, no fixation took place unless the crop was present. The stimulating effect of the crop is probably due to evolution of CO_2 from living and also from decaying roots. Further stimulation of the algae is afforded by the addition of P_2O_5 in either soluble or insoluble form, with or without K_2SO_4 or $MgSO_4$ but the two last applied singly have a depressing effect.

183.
DE, S. and CHANDRA, A.L. (1979) Degradation of organic nitrogenous wastes by a soil streptomycete.

Folia Microbiol. 24 : 473-477

A soil streptomycete which degraded hair, silk, wool feathers and leather was characterised and designated Streptomyces sp. A_9 56. It degraded leather to the greatest extent and solubilized 35.9% of the total nitrogen; 2.32 mg of glycine equivalent amino nitrogen could be obtained by degradation of 100 mg leather in 8 days.

184.
DEBNATH, N.C. and HAJRA, J.N. (1972) Transformation of organic matter in soil in relation to mineralisation of carbon and nutrient availability.

J. Indian Soc. Soil Sci. 20 : 95-102.

Three organic matters, viz., F.Y.M. (Farmyard manure), dhaincha (Sesbania), and paddy straw were incorporated in soil and incubated for 120 days to study the rate of their decomposition, pattern of nutrient mineralisation and their efficiency as soil organic matter builder.

The rate of CO_2 evolution from all the treatments was found to be highest within the first 48 hours of incubation. Farmyard manure has been graded as the best material amongst the three as soil organic matter builder, followed by dhaincha, straw being least efficient in this respect. As supplier of N, P and K to the immediate next crop, the efficiencies of the organic materials have been graded as follows: for N, dhaincha >F.Y.M. > Straw; for P, F.Y.M. > dhaincha > straw; and for k, straw > dhaincha > F.Y.M. Straw maintained nitrate depression throughout the period of incubation. After 120 days of incubation, both straw and F.Y.M. maintained higher soil pH than control, whereas soil with dhaincha maintained the same pH as that of control.

185.
DESAI, S.V. and FAZAL-UD-DIN (1937) The influence of non-nitrifying organisms on nitrification. Indian J. agric. Sci. 7 : 895-905.

In omeliansky solution soluble carbohydrates were decomposed by non-nitrifying organisms. Soluble organic constituents of soil and fermented manure did not depress nitrification. Nitrifying organisms live in symbiosis with non-nitrifying organisms, and are thereby able to overcome the toxic effect of soluble organic matter.

186.
DESHMUKH, V.A. and SHRIKHANDE, J.G. (1974) Effect of some herbicides on soil microflora.

J. Indian Soc. Soil Sci. 22 : 300-303.

Bacterial population in the red clay loam soil during the first week of incubation was depressed with all herbicidal treatments whereas a similar depression was noticed in the black clayey soil with only 2,4-D (H). Simazine (H) and Bladex. 2,4-D (L) and NaTA (L) stimulated bacteria during this period in the heavy soil. During the second week, only Simazine and Bladex could depress bacteria in the red soil and Simazine (H) and Bladex in the black soil. Actinomycetes were stimulated in presence of herbicides during the entire incubation period. Herbicides depressed growth of fungi in the light soil during the first week whereas a similar depression in the black soil was effected by 2,4-D (H), NaTA (H) and Bladex. Subsequently, marked depression of fungi in presence of herbicides was not observed.

187.
DESHPANDE, R.S.; KULKARNI, D.S. and VISWANATH, S. (1971) Influence of lime and nitrogenous fertilizers on soil biology.

Mysore J. agric. Sci. 5 : 77-81.

Limed and unlimed samples of acid red soil were incubated for up to 9 weeks with 0 and 500 ppm N as urea or $(NH_4)_2SO_4$, $(NH_4)_2SO_4$ in limed and unlimed soils decreased bacterial populations to levels below those of the control. Lime and urea, alone and together increased bacterial populations. NO_3-N levels were 36, 278, 286, 60, 310 and 354 ppm after 9 weeks' incubation in untreated and $(NH_4)_2SO_4$, urea, lime, lime+ $(NH_4)_2SO_4$ and lime + urea treated soils, respectively.

188.
DEVI. N.A.; KUTTY, R.K.; VASANTHARAJAN, V.N. and RAO, P.V.S. (1975) Microbial Metabolism of phenolic amines: degradation of dl-synephrine by an unidentified Arthrobacter.

J. Bacteriol. 122 : 866-873.

Microorganisms capable of degrading dl-synephrine were isolated from soil of Citrus gardens by enrichment culture, with dl--synephrine as the sole source of carbon and nitrogen. An organism which appears to be an arthrobacter, but which cannot be identified with any of the presently recognized species was predominant in these isolates.

189.
DEY, B.K. and BHATTACHARYYA, S. (1984) Influence of soil depth on growth and activity of Azotobacter chroococcum.

J. Indian Soc. Soil. 32 : 261-265.

Three strains of Azotobacter chroococcum in slants and broth culture were incubated at soil surface and at different depths in the same field from where the strains had been isolated. The growth characteristics (elevation, spreading and cell mass) and pigmentation of the strains in slants were better expressed in the 5 to 15 cm soil depth than at the surface, followed by a decline at still lower depths. The depth of the soil exerted a differential influence on the proliferation and efficacy towards nitrogen fixation of the strains in both culture. In general, under favourable nutritional, moisture and pH conditions, nitrogen fixation was the highest at 10 cm depth followed by that at 20 cm and the lowest at 30 cm depth in the soil.

190.
DHAR, B.; SINGH, B.D.; SINGH, R.B.; SRIVASTAVA, J.S. and SINGH, R.M. (1980) Seasonal incidence of rhizobiophages in soils around Varanasi.

Indian J. exp. Biol. 18 : 1168-1170.

Seasonal distribution of rhizobiophages of Rhizobium leguminosarum strain P5 was monitored in soil samples from 5 fields of diverse cropping history. Phages were detectable in all the fields although they were most numerous under legumes. The titre was very low during summer (May-July) and relatively high during winter (October-February), varying in parallel with fluctuations in Rhizobium populations. In general, phages isolated during winter showed a much faster thermal inactivation than those isolated during summer.

191.
DHAR, B.; SINGH, B.D., SINGH, R.B.; SRIVASTAVA, J.S.; SINGH, V.P. and SINGH, R.M. (1979) Occurrence and distribution of rhizobiophages in Indian soils.

Acta Microbiol. Polonica 28 : 319-325.

Soil samples from 30 fields in neighbouring districts of Varanasi were screened for rhizobiophages on 60 Rhizobium strains. Plaques were observed on five strains : P 1, p 5, SU 391 (R. leguminosarum). CB756 and 32H1 (Rhizobium sp.). Rhizobiophages infective on one or more of the five strains were present in all fields. There seems to be no correlation between presence of phage and the standing crop or the pH (7.1-8.2) of the soil. Eight distinct rhizobiophages have been isolated and characterized for host specificity, plaque morphology and maximum titer in broth.

192.
DHAR, B.; SINGH, B.D.;SINGH, R.M.; SINGH, V.P. and SINGH, R.B. (1980) Isolation and characterization of antibiotic resistant mutants of cowpea Rhizobium 32H1.

Indian J. exp. Biol. 18 : 1392-1395.

MNNG (N-methyl-N'-nitro-N-nitrosoguanidine) inactivation and spontaneous and MNNG induced mutation frequencies for resistance to streptomycin and erythromycin were studied in 32H1, a cowpea Rhizobium. The resistant mutants differed from the parent in growth pattern, phage sensitivity, infectivity of efficiency. The level of resistance of streptomycin resistant mutants was much higher in CS 7 medium than in yeast-mannitol medium.

193.
DHAR, N.R. (1943) Improvements of nitrogen status of soils and the origin of soil nitrogen. Nature 151 : 590-592.

A review. Carbonaceous substances assist the fixation of atmospheric N, particularly in light, and diminish loss of N by nitrification when ammonium sulphate is added to the soil. In light, N fixation can take place in the absence of bacteria. The behaviour of Indian soil is discussed, and the advantages of FYM are emphasised.

194.
DHAR, N.R. (1954) The role of organic matter in soil fertility. Lantbrttogsk Ann. 21 : 105-160.

A discussion with 75 references, mainly on the effect of organic matter on N-fixation. Conclusion reached concerning organic manuring and soil fertility are summarized in 31 points.

195.
DHAR, N.R. (1955) The role of organic matter in soil fertility. Ann. agron. Paris. 6 : 133-160.

A discussion mainly of the relationships between N-fixation and organic matter supply, including data on N losses in cultivated soils and on the better utilization of N from organic than from mineral fertilizers.

196.
DHAR, N.R. and BANERJEE, A.K. (1961) Fixation of atmospheric nitrogen in alkali soil by organic matter and phosphate. Proc. nat. Acad. Sci. India 31A : 309-311.

The increase in the efficiency of N fixation in an alkali soil (pH 10.32) on addition of rock phosphate was greater where paddy straw and especially Napier grass were incorporated in the soil, possibly due to the formation of nucleo-proteins and phosphoproteins that are more resistant than other proteins to bacterial decomposition.

197.
DHAR, N.R. and KAPOOR, R.C. (1948) Influence of temperature on nitrogen fixation in sterile and unsterile conditions.

Proc. Nat. Acad. Sci. India 17A : 57-64.

C oxidation and N fixation were greater in light at any temperature than in the dark. At 7° there was no fixation. The optimum temperature for C oxidation and N fixation in the dark was 35°. Differences between sterile and unsterile conditions were slight. The energy required for N fixation was liberated by C oxidation.

198.
DHAR, N.R. and PANT, M.C. (1949) Loss of nitrogen in bacterial nitrification. J. Indian Chem. Soc. 26 : 227-230.

During bacterial nitrification approximately half of the total N of hippuric acid, oil cake, uric acid, creatine and fibrin is lost. During nitrification, ammonium nitrite is produced as an intermediate product which decomposes into water and N. In soils with pH greated than 7, some N may be lost as ammonia gas, but under ordinary conditions, on adding N manures, acidity increases and the loss may be partially to the decomposition of nitrous acid. The presence of carbonaceous matter retards the rate of oxidation of ammonia to nitrate by acting as a negative catalyst.

199.
DHAR, N.R. and PANT, N.N. (1944) Nitrogen loss from soils and oxide surfaces. Nature 153 : 115-116.

In isolated flakes containing soil or a mixture of titanium and zinc oxides with urea or gelatine, there were losses of nitrogen and of carbon. These losses were smaller in flakes coverd with black cloth and the percentage loss was reduced by presence of glucose. It is concluded that non-biological denitrification is possible and that sunlight markedly accelerates the process.

200.
DHAR, N.R. and SESHACHARYULU, E.V. (1939) Nitrogen fixation and conservation in the soil. III-Effect of light on bacterial numbers and nitrogen fixation. J. Indian Chem. Soc. 16 : 463-476.

Exposure to sunlight increases N fixation in soil although it greatly decreases the contents of *Azotobacter*, total bacteria, and fungi. The increased N fixation is attributed to utilization of solar energy.

201.
DHAR, N.R. and SESHACHARYULU, E.V. (1939) Nitrogen fixation in soil not wholly a bacterial process. J. Indian Chem. Soc. 16 : 557-562.

Nitrogen fixation occurs under sterile conditions and is associated not only with bacterial activity, but also with the energy released during the oxidation of the organic substances.

202.
DHAR, N.R. and SESHACHARYULU, E.V. (1941) Nitrogen fixation under sterile conditions. Proc. nat. Acad. Sci. India 11 : 97-105.

With numerous carbohydrates it has been found that even under completely sterile conditions the N fixed per gram of carbon oxidized in light is appreciably greater than in the dark. It appears that the efficiency of N fixation is much the same whether *Azotobacter* is present or the soil sterile. The amount of N fixed both in the light and in the dark is increased in presence of certain metal oxides. It is suggested that the chief source of soil N is atmospheric N_2 fixed by a surface reaction.

203.
DHAR, N.R.; BOSE, S.M. and SINHA, S.B. (1952) Efficiency of different methods of nitrogen fixation. Proc. nat. Acad. Sci. India 21A : 137-155.

Experiments involving illumination by a 60-watt electric bulb of Ca- and P- rich soils mixed with straw showed a fixation of up to 150 mg of N per gram of oxidized C under unsterile condition. The fixation of N was less in the dark or under sterile conditions. It is suggested that the energy of oxidation is utilized in breaking up water molecules, and the atomic H into NH_4. The N fixation in light is said to be as high as that obtained throught symbiotic organisms.

204.
DHAR, N.R.; CHATTERJI, B.G. and PANT, N.N. (1952) The relative influence of surface and bacterial oxidations in the fixation of nitrogen. Proc. nat. Acad. Sci. India 21A : 126-136.

Experimental results show that N-fixation takes place when soil or insoluble substances like Fe_2O_3, ZnO, Al_2O_3 etc. are mixed with energy materials like dung, leaves, starch and allowed to undergo slow oxidation in air under sterile and unsterile conditions. The N-fixation in sunlight is greater than in the dark. The surface oxidation of these materials is more pronounced than microbial oxidation.

205.
DHAR, N.R.; SESHACHARYULU, E.V. and BISWAS, N.N. (1941) New aspects of nitrogen fixation and loss in soils.

Proc. Natn. Inst. Sci. India 7 : 115-131.

Nitrogen is fixed non-bacterially (more in the light than in the dark) during the oxidation of organic matter in the soil.

206.
DHAR, N.R.; SESHACHARYULU, E.V. and MUKERJI, S.K. (1937) Photochemistry of nitrogen fixation and resemblance between nitrogen fixation and carbon assimilation. J. Chem. Phys. 34 : 756-763.

Field and green house trials indicated that N-fixation is greater in light than in darkness, and that light is utilized for it in the same way as in carbon assimilation. The required energy can also be provided by the oxidation of energy rich material.

207.
DHAR, N.R.; SESHACHARYULU, E.V. and MUKERJI, S.K. (1937)

Similarity between nitrogen fixation and carbon assimilation. Proc. nat. Acad. Sci. India 7 : 140-148.

Experimental results from laboratory and field show that N fixation per gram of C oxidized in sunlight or artificial light is much greater than that in the dark. Energy required for N fixation can be supplied from the oxidation of energy materials and from light.

208.

DHARMAJI RAO, S. and PATNAIK, H. (1978) Effect of blue--green algae on the vegetative growth of two early varieties of rice. pp. 210-214. In S.P. Sen, Y.P. Abrol and S.K. Sinha (eds.), Nitrogen Assimilation and Crop Productivity, Associated Publishing Company, New Delhi.

The effect of two blue-green algae, Nostoc commune and Oscillatoria tenuis on the vegetative growth of two early varieties of rice. Taichung native-I and ptb-10 was studied. The introduction of heterocystous blue-green algae N. commune in both the early rice varieties enhanced vegetative growth than that of the non-heterocystous blue-green algal, O. tenuis. The early rice varieties ptb-10 produced more tillers and leaves in the presence of algae plus ammonium sulphate. The application of non-heterocystous blue-green alga O. tenuis markedly increased the number of tillers, leaves and plant height as against control in the rice variety ptb-10

209.

DUBE, J.N.; NAMDEO, S.L and JOHAR, M.S. (1975) Coal as a carrier of rhizobia.

Curr. Sci. 44 : 434.

Inferior grade coal dust appeared to be a promising carrier for R. japonicum.

210.

DUBE, V.P.; CHARAYA, M.U.; MODI, P. (1980) Ecological and in vitro studies on the soil mycoflora of mango orchards.

Proc. Indian Acad. Sci. 89 : 151-160.

A study of the soil mycoflora of mango orchards over a 10 month period revealed a significant correlation between soil moisture and fungal pupulations. On the other hand, soil reaction, organic matter, soil temperature and conductivity showed no significant correlation with fungal population. Marked variations in the fungal flora from month to month were observed. Aspergillus flavus, Aspergillus niger and Cladosporium herbarum showed the capacity to decompose mango leaves in vitro. Small concentrations of mango leaf powder had a stimulating effect on the mycelial growth as well as sporulation of these species. These factors perhaps contribute to the dominance of the three fungal species in mango orchards.

211.
DUTTA, B.G. and GHOSH, G.R. (1964) Soil fungi from Orissa (India). IV. Soil fungi of paddy fields.

Mycopath. Mycol. appl. 25 : 316-322.

A hundred and sixty-five fungal species of 51 genera were isolated from paddy fields at different stages of cultivation. Some species were markedly dominant and persisted throughout the whole cycle.

212.
DUTTA, B.G.; GHOSH, G.R. and ORR, G.F. (1963) Soil fungi from Orissa (India). II - A new species of Pseudorachniotus. Mycologia. 55 : 775-780.

P. echinulatus isolated from a paddy soil, is described and illustrated.

213.
DUTTA, B.G.; GHOSH, G.R. and ORR, G.F. (1984) Soil fungi from Orissa (India). III - A new species of Pseudorachniotus with punctate spores. Mycologia 56 : 153-157.

A new species of P. punctatus is described and illustrated.

214.
DUTTA, N. and VENKATARAMAN, G.S. (1958) An exploratory study of the algae of some cultivated and uncultivated soils.

Indian J. Agron. 3 : 109 -115.

Samples from 0, 2, 6, and 12 inches from one uncultivated and three cultivated alkaline soils were examined for the presence of algae. Blue-green algae predominated while green algae and diatoms were few. Most of the blue-green algae were capable of fixing N.

215.
DUTTA, S. and CHAUDHURI, H. (1944) Fungus flora of the soil and reduction of carbohydrates by them. Proc. Indian Sci. Cong. 31 : 62.

A gradual increase in the species and numbers of the mycoflora from unreclaimed to reclaimed soils was registered in an ecological survey. The development of soil fungi is promoted by an abundance of organic nutrient and a slightly acid reaction, and is more copious in the winter than during the summer months.

216.
DWIVEDI, G.; MATHUR, R. and MATHUR, S.N. (1985) A comparative study of efficiency of various rhizobial strains on nodulation and N_2- fixation in mung-been (*Phaseolus aureus* cv. Pusa Baisakhi).

Proc. Natn. Sem. Plant Physiol. (Abst.) p. 38

Specificity of Rhizobium for mung-bean cv. Pusa Baisakhi has been worked out. Nodulation in terms of nodule number, their fresh and dry weights, and nitrogen fixation capacity have been worked out. Results showed strain No. 441 to be most suitable for *P. aureus* cv. Pusa Baisakhi.

217.
DWIVEDI, R. (1972) A technique for isolation of penicillia from soil.

Pl. Soil 37 : 171-172

A technique based on the inclusion of Nystatin in Czapek's medium is suggested.

218.
DWIVEDI, R. and DWIVEDI, R.S. (1972) Rhizosphere microflora of coriander with emphasis on fungistasis.

Ann Inst. Pasteur 122 : 455-461.

The rhizosphere of Coriandrum sativum contained fewer fungal species than the non-rhizosphere soil, probably due to the secretion of antifungal substances by the root. Tables of the frequency and abundance of rhizosphere and non-rhizosphere fungi are presented.

219.
DWIVEDI, R. and DWIVEDI, R.S. (1976) Effect of antibiotics on fungal colonization of agar discs from soil inocula.

Proc. Indian natn. Sci. Acad 42 : 205-208.

Effect of griseofulvin, polymyxin-B-sulphate, benzyl penicillin and Actidion in varying concentrations was studied on colonization of fungi on agar discs from mixed inocula of soil. As the concentrations of antibiotics increased, the number of fungal colonies and representative species decreased. In the highest concentration of each antibiotic the colonization of agar discs was restricted to only a few genera of fungi.

220.
DWIVEDI, R.S. and DWIVEDI, R. (1971) Ecology of soil fungi of a sal forest with emphasis on fungistasis.
Pl. Soil 34 : 33-42.

Samples of soil were collected from 3 forest regions of different age. Soil pH varied from 6.6 to 6.9 and in all samples the upper horizon contained more organic matter, N, P and other elements which favoured the growth of fungi. The fungal population was represented by 24 genera and 49 species and the number of fungi decreased with depth. The pattern of colonization of nutrient agar changed with the increase of concentration of staling products produced by earlier established fungal colonies. At the highest concentration of staling the colonizing species were Trichoderma, Aspergillus, Penicillium,

Curvularia, Alternaria and _Fusarium_. Fungistatic activity was higher in unsterilized soil.

221.
DWIVEDI, R.S. and PATHAK, S.P. (1978) Population dynamics of _Fusarium oxysporum, F. lycopersici_ in the fungicides amended soil.

Proc. Indian natn. Sci. Acad. B 43 : 207-212.

Autoclaved and unautoclaved field soil, artificially infested with _F. oxysporum_ f. _lycopersici_, was treated with 100 to 800 ppm (W/W) of PCNB, Dithane Z-78, Difolatan and Zincop. The population was recorded after 20, 50 and 80-day intervals. Population of the test fungus tolerated all PCNB concentrations under both soil conditions. Although Dithane Z-78 and Zincop were effective, they did not check the population completely even at the highest concentration used. The minimum population was recorded after a 20-day interval in natural soil. Difolatan was the most lethal of the fungicides tested, completely checking the fungal colony at 800 ppm under both soil conditions.

222.
DWIVEDI, R.S. and SHUKLA, A.N. (1977) Fungal decomposition in relation to carbon dioxide evolution in a tropical sal forest biome.

Proc. Indian natn. Sci. Acad. 43 : 26-32.

In a one-year experiment in which Sal litter was buried in nylon bags in the soil of a _Shorea robusta_ forest, Phycomycetes were the fungi initially isolated from the decomposing litter and were later replaced by cellulose-decomposing Ascomycetes and Deuteromycetes. A white felt of Basidiomycete mycelium frequently occurred in the litter, CO_2 release during decomposition was correlated with soil temperature and fungal population.

223.
DWIVEDI, R.S. and SINGH, K.P. (1972) Establishment of microfungi in the root region of _Andrographis paniculata_ Nees.

Proc. Indian natn. Sci. Acad. B 38 : 87-93.

Some of the fungi present in the rhizosphere and on the root surface of the medicinal plant A. paniculata decreased or increased according to the age of the plant; some species were confined to a particular stage. These isolation patterns indicated the colonising ability of the various species.

224.
DWIVEDI, R.S. and SINGH, U.P. (1974) Comparative cellulolytic ability of some litter-inhabiting fungi.

Proc. Indian natn. Sci. Acad. B 40 : 420-423.

The comparative cellulolytic ability of twenty species of fungi from the litter of Terminalia spp. was assayed in vitro. These fungi could be classified into three groups depending upon their cellulolytic ability, which was inversely related to hyphal dry-matter production.

225.
EMMIMATH, V.S and RANGASWAMI, G. (1971) Studies on the effect of heavy doses of nitrogenous fertilizer on the soil and rhizosphere microflora of rice.

Mysore J. agric. Sci. 5 : 39-58.

In plot experiments, rice growing in red soil was fertilized with 0, 40, 80 and 160 lb/acre N and 40 lb/acre each of P and K. N application significantly affected the numbers of bacteria, actinomycetes and fungi. There was a selective stimulation of bacteria and fungi in the rhizosphere during the early stages of plant growth but actinomycetes increased in both soil and rhizosphere during the later stages of growth. Fertilizing decreased Gram + ve and spore-forming bacteria. P and K alone increased actinomycetes and Azotobacter populations in the rhizosphere but high N rates inhibited Azotobacter.

226.
FOWLER, G.J. and KOTWAL, Y.N. (1924) Chemical factors in

denitrification. J. Indian Inst. Sci. 7 : 29-38.

The evolution and consequent loss of gaseous nitrogen taking place in nature, or in the operations of agriculture and sewage purification due to purely chemical causes is negligible, as far as the reactions investigated are concerned. The losses of nitrogen due to bio-chemical changes occur in a great variety of ways.

227.
GADRE, S.V.; MAZUMDAR, L. and MODI, V.V. (1967) Interspecific transformation in Rhizobium.

Arch. Mikrobiol. 57 : 388-391.

Two characters, "penicillin resistance and frustose-utilizing ability" "were transferred from a Rhizobium Ohaseoli culture (from French bean roots) to a Rhizobium of the cow-pea group (from groundnut roots); the transfer was effected by treating the recipient culture with DNA extracted from the donor.

228.
GANGAWANE, L.V. (1978) Tolerance of aureofungin and streptocycline by N-fixing blue-green algae.

Hindustan Antibiot. Bull. 21 : 36-37.

The blue-green algae Westiellopsis sp. (IARI156) Aulosira sp. (IARI168) Nostoc sp. (IARI 221), Tolypothrix sp. (IARI 76) and Calothrix sp. (IARI 91) were used in the study. Tolerance limit of aureofungin by these algae is more than that of the recommended dose for rice seed treatment. None of these organisms could grew in the medium containing streptocycline.

229.
GANGAWANE, L.V. (1978) Tolerance of N-fixing blue-green algae to Brassicol, Bavistin and Fytolan.

J. Indian Bot. Soc. 57 (suppl.) : 3

With the algal species tested (Westiellopsis sp., Aulosira sp., Nostoc sp., Tolypothrix sp. and Calothrix sp.) higher rates (500-1000 ppm) of Brassicol were tolerated than of Bavistin and Fytolan. Caolthrix was completely sensitive to Bavistin. The growth of Nostoc sp. and Tolypothrix sp. was stimulated by Bavistin at 100 and 500 ppm respectively. Tolypothrix sp. and Aulosira sp. were not affected by Brassicol or Fytolan. Germination of the seeds of three rice varieties (Suhasini, Surya and Satya) was not affected by these fungicides, except in a few cases where a 5-10% reduction in germination was observed. Bavistin stimulated the root and shoot growth of all varieties. Shoot length of Satya was promoted by all the fungicides.

230.
GANGAWANE, L.V (1979) Tolerance of Thimet by nitrogen fixing blue-green Algae. Pesticides 13 : 33-34

Thimet 10G had no harmful effects on Westiellopsis, Aulosira, Tolypothrix and Calothrix at concentration of 1000, 500, 500 and 300 ppm (a.i.) respectively, but Nostoc was unable to tolerate a concentration as low a 1 ppm.

231.
GANGAWANE, L.V. (1980) Tolerance of N-fixing blue-green algae to Brassicol, Bavistin and Fytolan. J. Indian Bot. Soc. 59 : 157-160.

Tolerance of five nitrogen-fixing algae (Westiellopsis sp., Aulosira sp., Nostoc sp., Tolypothrix sp., Calothrix sp.) to three fungicides was studied. All species could tolerate fungicides at some concentrations. Higher concentrations (500-1000 ppm) of Brassicol were tolerated than of Bavistin and Fytolan. Germination was little affected or unaffected by the fungicides. Root growth was stimulated by Bavistin.

232.
GANGAWANE, L.V. and DESHPANDE, K.B. (1978) Agronomic treatments and changes in the rhizosphere mycoflora of groundnut. VI-Effect of partial sterilization of soil.

Rev. Ecol. Biol. Soil 15 : 311-321.

Sterilization of soil with formaldehyde, acetic acid, BHC and blotix temporarily reduced the size of the fungal population of rhizosphere and non-rhizoshere soil. Equilibrium was regained within 30 days in the former, and within 60 days in the latter. Each chemical had a different qualitative effect on the composition of the fungal flora, different species being stimulated, inhibited or unaffected by different chemicals.

233.
GANGOPADHYAY, S.; WYLLIE, T.D. and TEAGUE, W.R. (1982) Effect of bulk density and moisture content of soil on the survival of Macrophomina phaseolina. Pl. Soil. 68 : 241-247

A gravimetric moisture content of w = 0.15 supported high sclerotia numbers whereas higher (w = 0.25) and lower (w = 0.05) moisture levels did not. High bulk density Db = 1.56 did not support fungal sclerotial numbers as well as Db = 1.30 or 1.15. Survival was poor when Db = 1.56 and w = 0.25 were combined.

234.
GANTOTTI, B.V. and RANGASWAMI, G. (1971) Improvement in the physical properties of soil under the influence of the rhizosphere microflora of four different plant species. Pl. Soil 35 : 347-356.

Qualitative studies on the edaphosphere microflora indicated the continuance of rhizosphere effect beyond the root surface region. The rhizosphere microflora in association with roots of the growing plants is suggested to play a pivotal role in improving soil structure.

235.
GARG, F.C.; TAURO, P.; MISHRA, M.M and GROVER, R.K. (1981) Effect of fungicides on growth and on DNA, RNA and protein anabolism of a Rhizobium sp.

Ann. Microbiol. 132 B : 231-239.

Eleven fungicides were examined for their effect on the growth and anabolism of Rhizobium sp. in pure culture. Only 4 of the fungicides tested were found to inhibit growth of rhizobia at a concentration as low as 10ug/ml. Growth inhibition by these fungicides appeared to be primarily due to inhibition of respiration, although thiophanate and anilazine also affected DNA and RNA synthesis, respectively. Inhibition was, however, temporary and the bacterium had the ability to overcome this initial inhibition without mutation.

236.
GARG, R.P.; NAIR, S.K. and DADARWAL, K.R. (1983) Associative nitrogen fixing bacteria of bajra (Pennisetum americanum).

Curr. Sci. 52 : 138-140.

In a survey of bacteria Klebsiella sp. and Enterobacter sp. were found to by major taxonomic types associated with bajra.

237.
GAUR, A.C. and MISHRA, K.C. (1978) Dynamics of microbial population in soil as influenced by simazine and ecological factors.

Zentbl. Bakt. ParasitKde Abt. II 133 : 357-361.

Simazine at 0, 2, 20 and 200 ppm was added to samples of alluvial soil which were incubated for 10 weeks Bacteria and fungi were inhibited to a greater extent than actinomycetes; Azotobacter and actinomycetes were stimulated by 2 ppm. Simazine was less toxic to bacteria and fungi at pH 5 and 9.6 than at 7.4, and less at 15 C than at 30 or 45 C. A. high moisture regime reduced the toxicity of simazine to actinomycetes and fungi, and incorporation of 2% farmyard manure reduced its toxicity to fungi.

238.
GAUR, A.C. and PAREEK, R.P. (1971) Tolerance of nitrification to DDT.

Indian J. Entom. 33 : 368-370.

The effect of DDT on the nitrification of $(NH_4)_2SO_4$ in an alluvial sandy loam soil was studied in laboratory experiments. DDT did not affect the nitrification rate, but did reduce the numbers of Nitrosomonas and Nitrobacter in the soil when applied at rates @ 5 and @ 100 ppm respectively.

239.

GAUR, A.C.; ARORA, D. and PRAKASH, N. (1979) Electron microscopy of some rock phosphate dissolving bacteria and fungi.

Folia Microbiol. 24 : 314-317.

Pseudomonas striata, Bacillus polymyza, B. megaterium, B. Pulvifaciens, Aspergillus awamori, A. niger and Penicillium digitatum dissolved tricalcium phosphate much more than they dissolved Mussorie and Udaipur rock phosphate. Solubilization was greatest by A awamori, A. niger and P. striata. An electron-dense layer was revealed on the bacterial surfaces by means of negative staining. Incubation for 3 weeks decreased the size of the tricalcium phosphate particles from 140-250 to 30-9 0nm.

240.

GAUR, A.C.; MATHUR, R.S. and KAVIMANDAN, S.K. (1969) In vitro nitrogen fixation by Azotobacter as influenced by plant materials and their extracts.

Agrochimica 13 : 81-84.

Aqueous extracts of berseem and barley increased N fixation by Azotobacter, but residues of the plants added as such to the culture medium decreased N fixation.

241.

GAUR, A.C.; SADASIVAM, K.V. and VIMAL, O.P. (1971) A study on the decomposition of organic matter in an alluvial soil : CO_2 evolution, microbiological and chemical transformations.

Pl. Soil 35 : 17-28.

Berseem hay, farmyard manure (FYM) and wheat straw were incubated with an alluvial sandy loam at 30°C. Rate of decomposition and effect on soil microflora were studied. Microbial activity varied with time of incubation, with maxima at 30 and 90 days and minima at 60 and 120 days. Wheat straw stimulated Azotobacter and FYM increased the fungal population. The different treatments variously increased the organic matter and nutrient status of the soil after incubation.

242.
GAUR, Y.D.; SEN, A.K. and SUBBA RAO, N.S. (1974) Problem regarding groundnut (Arachis hypogaea L.) inoculation in Tropics with special reference to India. Proc. Indian natn. Sci. Acad. 40B : 562-570.

Despite the fact, that the area under groundnut cultivation in India is the highest in world, the yields are relatively poor and amounts to only one third of the yield obtained in Europe. The roots of groundnut are nodulated by Rhizobium species of 'Cowpea' group which is typically tropical in nature. Although the native rhizobia profusely nodulate the cultivated varieties, artificial inoculation of seeds with laboratory grown cultures has not increased the yield of groundnut as shown by the results of field trials done in different parts of India. The present paper discusses the factors responsible for the failure to obtain the advantages of artificial inoculation with Rhizobium in so far as this important oil yielding, protein-rich crop is concerned. The paper highlights the future line of work that may be undertaken to improve yields through Rhizobium inoculation.

243.
GAUR, Y.D. and SUBBA RAO, N.S. (1984) The need for Rhizobium phaseoli inoculation to establish American beans in India.

Curr. Sci. 53 : 430-431.

Some of the American beans introduced in India are unsuccessfully cultivated during the cooler part of the year in plains and in the warmer part of the year in the hills. The results showed that nodulation was virtually absent in plants grown in all the soils except for stray occurrence of ineffective tiny nodules on roots due to the promiscuous nature of indigenous strains of rhizobia.

244.
GAUR, Y.D.; SEN, A.N. and RAO, N.S.S. (1980) Improved legume Rhizobium symbiosis by inoculating preceding cereal crop with Rhizobium.

Pl. Soil. 54 : 313-316.

In a cereal/legume crop rotation system, better Rhizobium symbiosis was obtained by double inoculation when the preceding cereal crop of maize was also inoculated with the same Rhizobium strian used to inoculate the following legume crop of green gram (Vigna radiata) or groundnut.

245.
GAUTOM, R.; DATTA, T.; SOOD, M.L. and KALRA, I.S. (1984) Studies of bacterial population of soil under different levels of fertilizer application.

J. Indian Soc. Soil. Sci. 32 : 178-179.

Effect of fertilizers and FYM application on soil bacteria was investigated in a field experiment on a sandy soil. Application of N, P and K at normal dose with FYM increased the number of bacteria; N, P and k singly was less favourable and the effect of N was greater than that of P and K.

246.
GHAI, J.; GHAI, S.K. and KALRA, M.S. (1983) UV-radiation induced changes in antibiotic markers, chemical composition of water soluble polysaccharides and nodulation ability of Rhizobium trifolii 11B. Indian J. exp. Biol. 21 : 500-503.
Studies with R. Trifolii strain 11B and nine UV-radiation induced mutants revealed that there was a direct correlation between nodulation ability (of Trifolium alexandrinum, the host of 11B) by the rhizobia and their ability to produce water soluble

polysaccharides (WSP), those producing the greatest amounts of WSP inducing the greatest number of nodules.

247.
GHILDYAL, B.P. (1963) Influence of tropical season on carbon and nitrogen transformations; nitrite and nitrate formers and bacterial plate count. Soil Sci. Plant Nutr. 9(2) : 28-34.

In soil of the humid sub-tropical alluvial plain, bacterial numbers decreased sharply at the onset of rain and subsequently increased in the rainy season until winter. Appreciable numbers of NO_2 and NO_3 formers were present only in summer. Even at high soil temperature and low soil moisture the soil micro-organisms remained active.

248.
GHILDYAL, B.P. (1964) Some observations on the nature of non-biological nitrification in tropics.

J. Indian Soc. Soil Sci. 12 : 51-57.

Influence of solar heat of tropics on the oxidation of ammonium sulphate in sterilized distilled water land sand is reported. About 2 to 90 per cent of the ammoniacal nitrogen introduced was oxidized to nitrate. The nitrification was accompanied by high nitrogen losses varying from 34 to 68 per cent which may be as nitrous or nitric oxide. Temperature of cultures, length of exposure to sunlight, season, and initial nitrogen content influence the non-biological nitrification and cayalytic oxidation of ammonium sulphate.

249.
GHOSH, S.R. (1979) A study of the relative efficiency of mustard oil cake and urea and the effect of salinity in plankton production in Indian fishponds. Hydrobiologia 62 : 89-92.

In laboratory trials Glosterium, Fragilaria, Pinnularia and Cyrosigma grew well on saturated soil media in the salinity range 24 to 30 ppt, whereas lower salinities ranging from a trace to 8 ppt were more suitable for production of Anabaena ,

Synedra, Navicular, Amphipleura, Amphora and Nitzchia. Plankton production was greater on media amended with urea than with mustard oil cake as a nitrogen source.

250.
GOPALA RAO, G. and DHAR, N.R. (1934) Photolysis of amino acids in sunlight. J. Indian Chem. Soc. 11 : 617-622.

Decomposition of amino acids in sunlight suggests that ammonification can take place in soils exposed to sunlight in the absence of bacteria or fungi.

251.
GOPALKRISHANAN, S.; RAJA, P.N. and RAJAGOPALAN, N. (1964) Mineral nutrition and nitrogen fixation in groundnut. Curr. Sci. 33 : 391-393.

Results suggest that the depressing effects of high nitrate levels on N fixation and on haemoglobin content of nodules may be offset by an optimal supply of Mg and Fe to the symbiotically growing plants.

252.
GOSWAMI, K.P. and PAREEK, R.P. (1978) Nitrogen economy by leguminous crops: Nitrogen balance in soyabeen - wheat and gram. pp 14-26. In S.P. Sen, Y.P. Abrol and S.K. Sinha (eds.), Nitrogen Assimilation and Crop Productivity. Associated Publishing Company, New Delhi.

Quantitative data on N_2-fixation by legumes vary a great deal. Field studies with soyabean have shown profuse nodulation following inoculation with five strains of R. japonicum which subsequently gave upto 133% increase in soyabean yields. The highest amount of dinitrogen fixed by strain 1028 was estimated to be 256 kg ha-1. The other strains which gave nearly 100% increase in yields were 1005, 1020 and Mixed-Strain which fixed 187,89 and 133 kg N ha-1. The residual fertility effect of soyabean on wheat crop was determined; plots treated with 1020 gave significantly higher wheat yield (54.4 and 16.3% increase with 20 and 60 kg N ha-1, respectively) followed by 1005 and 1020. The wheat yield was above 50 Q ha-1 with only 60 kg N applied, out of which 5 to 7 Q was due to the residual effect of soyabean inoculation. Eleven strains of rhizobia were evaluated with respect to gram under field conditions. The most significant increase viz. 32% in yield was obtained with the strain 6051. Straw mulching with inoculation gave 57% increase in gram yield. The nitrogenase activity of various strains were measured (in excised roots of gram with intact nodules) at different stages of growth by the acetylene reduction technique. The measured nitrogenase activity did not follow strictly the pattern of yield response by the different strains of rhizobia. The study supports the convention that substantial saving of fertilizer-N can be made by including legumes in rotation.

253.
GOSWAMI, K.P. and SEN, A. (1962). Solubilization of calcium phosphate by three strains of a phosphoba- cterium. Indian J. agric. Sci. 32 : 96-101.

Three strains of Bacillus megatherium var. phosphaticum were obtained from (a) phosphobacterin (b) Fosfo 24 L and (c) from the glands of Cassia occidentalis. Maximum solubilization of PO_4 occurred within 2 weeks under cultural conditions. With peptone as N source (b) dissolved more PO_4 than (a) did. With $NaCO_3$, (a) dissolved more PO_4 than did (b).

254.
GOVINDARAJAN, K. and KANNAIYAN, S. (1983) Influence of Azolla inoculation on rice crop with and without nitrogen application.

Indian J. Agron. 28 : 178-179.

The results showed that the incorporation of Azolla significantly increased the grain yield. The yield from the plots treated with 75 kg N/ha along with Azolla was on par with 100 kg N/ha. Similarly, Azolla was on par with 25 kg N/ha.

255.
GOYAL, S.K. (1983) Diazotrophic algae and soil fertility. Proc. Natl. Symp. Adv. Front. Pl. Sci. Jodhpur 9 : 99.

Under tropical rice field conditions, blue-green algae from the most promising source of biologically fixed nitrogen, contributing 25-30 kg N/ha season. In addition to nitrogen fixation, they have also been shown to liberate biologically potent substances like vitamin B12 and auxins, solubilize insoluble phosphates, increase nitrogen availability to the crop and improve the physical and chemical properties of the soil. They also bring about an addition to the organic matter content of the soil.

256.
GOYAL, V.; CHETAL S. and NAINAWATEE, H.S. (1984) Water stress alters characteristics of alkaline phosphate in *Rhizobium* and *Azotobacter* species. Folia Microbiol. 29 : 228-232.
Rhizobium trifolii T 19C and *Azotobacter chroococcum* B 13 were grown under polyethylene glycol (6000) induced water stress in the liquid media. Water stress at -200 and -300 kPa resulted in a several-fold increase in the activity of alkaline phosphatase in both bacteria. Optimum pH and temperature and thermostability of the enzyme were also affected by the water stress.

257.
GRAHAM, P.H. (1969) Selective medium for growth of *Rhizobium*. Appl. Microbiol. 17 : 769-770.

Of 52 Rhizobium strains, all but three grew well on a yeast-mannitol-antibiotic-agar medium. None of the few Agrobacterium strains tested grew well on it. The medium gave 37.5-9.3% recovery of Rhizobium strains added to soil.

258.
GUHA SIRCAR, S.S.; DE, S.C. and BHOWMICK, H.D. (1940) Microbiologiclal decomposition of plant materials. II - changes in the methoxyl and nitrogen content of lignin of rice straw during

its decomposition by microorganisms. Indian J. agric. Sci. 10 : 152-157.

There is an increase in methoxyl content of lignin with the process of decomposition. Loss of methoxyl is greatest under anaerobic, intermediate under waterlogged and least under aerobic conditions. The reverse is the case with increment of N content,

259.
GULATI, S.L. and SUBBA RAO, N.S. (1973) Effect of additional shots on growth of Rhizobium trifolii in relation to mass cultivation of Rhizobia.

Hindustan Antibiot. Bull. 16 : 84-88.

Addition of case in hydrolysate, berseem seed extract and yeast extract enhanced the growth of a strian of Rhizobium trifolii on yeast extract mannitol basal medium whereas starch, calcium carbonate, phosphoric acid and citric acid depressed growth. The stimulation of growth was more pronounced when substances were added, just after lag phase even at low concentrations.

260.
GUPTA, B.R. and BAJPAI, P.D. (1974) Some microbiological studies in salt-affected soils I-Pattern of soil microbial population as affected by salinity and alkalinity.

J. Indian Soc. Soil Sci. 22 : 176-180.

Manimum number of bacteria, fungi, Azotobacter and nitrifiers were observed at neutral pH and normal salinity. Proportion of actinomycetes to bacteria was higher in moderately to highly salt affected than normal soils. Nitrite formers were higher and were possibly less salt sensitive than nitrate former. A considerable decrease in population of bacteria, fungi, Azotobacter, was noticed with increase in ECe and pH and appeared to be greatly the function of the latter. Bacterial spores were more at higher salinity levels.

261.
GUPTA, B.R. and BAJPAI, P.D. (1974) Some microbiological studies in salt-affected soils. I-Pattern of soil microbial population as affected by salinity and alkalinity. J. Indian Soc. Soil Sci. 22 : 176-180

Maximum number of bacteria, fungi, Azotobacter and nitrifiers were observed at neutral pH and normal salinity. Proportion of actinomycetes to bacteria was higher in moderately to highly salt affected than normal soils. Nitrite formers were higher and were possibly less salt sensitive than nitrate fromer. A considerably decrease in population of bacteria, fungi, Azotobacter, was noticed with increase in ECe and pH and appeared to be greatly the function of the later. Bacterial spores were more at higher salinity levels.

262.
GUPTA, B.R. and BAJPAI, P.D. (1978) Some microbiological studies in salt affected soils. III-The effect of some inorganic ameliorants on nitrogen mineralization. J. Indian Soc. Soil. Sci. 26 : 173-178.

In vitro, application of inorganic ameliorants as gypsum sulphuric acid and calcium chloride in salinealkali soils increased the quantum of minerlized nitroggen products as compared to control. In unreclaimed and partially reclaimed soils, nitrite-N was observed to be accumulated. Ammonification rate was observed outstripping that of nitrification. At the initial stage, the differences in contents of available nitrogen between control and amended soils were low but they became larger as the days of incubation prolonged. Ammonical-N showed a fluctuating trend during the entire period of incubation on while nitrate-N showed accumlation up to 75-90 days in all the cases. Mineralization process was higher under flushed condition than in unflushed ones.

263.
GUPTA, K.G. and SEN, A. (1962) Aggregation of soil due to growth of Rhizobium sp. from some cultivated legumes. Soil Sci. 94 : 345-348.

It was observed that highest aggregation of the soil was caused by Rhizobium sp. from pea. On the average, the amount of aggregation caused by the growth of the isolates came to 59 per cent in the case of Rhizobium sp. from gram, 41 per cent in the case of methi and 83 per cent in the case of isolates of Rhizobium sp. from pea.

The amount of aggregation of soil could not be correlated with the amount of gum produced by the isolates under cultural condition or with their growth in the soil.

264.
GUPTA, K.G. and SEN, A. (1963) Utilization of combined nitrogen by Rhizobium sp. from some common cultivated legumes in relation to their efficiencies. Indian J. agric. Sci. 33 : 240-243.

Utilization of asparagine-N by Rhizobium isolates from a single legume species varied considerable. With isolates from pea, fenugreek and black gram there was a positive and significant correlation between amounts of N utilized and effectiveness of N-fixation.

265.
GUPTA, K.G. and SEN, A. (1965) The relationship between glucose consumption by Rhizobium spp. from some common cultivated legumes and their efficiencies. Pl. Soil 21 : 229-238.

Increase in efficiency per unit of glucose in rhizobia from black gram and fenugreek was about seven times that in strains from pea. This relationship was connected with seed size, been seeds being largest and fenugreek seeds smallest.

266.
GUPTA, K.G. and SEN, A . (1965) Relationship between nitrogen fixing efficiency of Rhizobium spp. and their capacities of glucose consumption and phosphate utilization.

Indian J. agric. Sci 35 : 39-42.

The efficiency of the different isolates of a single Rhizobium sp. was related positively and significantly to their capacity to consume glucose and utilize P. Capacity to consume glucose was a more accurate index of efficiency than capacity to utilize P.

267.
GUPTA, K.G and SEN, A. (1966) Utilization of phosphates by Rhizobium spp in relation to their efficiencies.

Soil Sci. Pl. Nutr. 12 : 18-22.

There was a marked variation in the amounts of P utilized by strains from a single legume. From a positive and highly significant correlation between the amounts of P utilized and efficiencies of the isolates, it could be inferred that strains of high efficiency utilized large amounts of P.

268.
GUPTA, K.G. and SHIRKOT, C.K. (1981) Development of resistance in slow-growing rhizobia to 1-naphthyl- -N-methyl carbamate (sevin).

Pl. Soil 60 : 399-407.

Three strains of rhizobia (Rhizobium melilotti and strains isolated from Lotus corniculatus and Vigna unguiculata), resistant to 60, 70, or 50/ug ml-1, respectively, of sevin, were developed by serial culture in media containing increasing concentration of the pesticide. The resistance was stable during 20 transfers in pesticide free medium. Resistant strains showed decresed growth rates in comparison with sensitive strains. Dehydrogenase activity was higher in two resistant strains and lower in the third than in their sensitve counterparts. The development of resistance resulted in decreased lipid contents and altered lipid composition.

269.
GUPTA, K.G; AGGARWAL, J.C. and MAKKAR, N.S. (1974) Sensitivity or rhizobia to vanillin.

Folia Microbiol. 19 : 317-321.

Rhizobia isolated from gram (Cicer arietinum) showed marked resistanece to vanillin, while those isolated from fenugreek, pea, cowpea and urid (Phaseolus mungo), together with R. melioti and R. trifolii, were sensitive; R. japonicum was intermediate. When grown in broth culture and in soil, manure and peat, resistant strains were able to utlize vanillin as the sole C source. Sensitivity was in some cases reduced by the addition of mannitol.

270.
GUPTA, K.J.; SANDHU, J.S. and SUD, R.K. (1971) Another approach on the artificial inoculation of legumes.

Ann. Inst Pasteur, 120 : 665-670.

In pot experiment, Rhizobium spp., specific for Vigna sinensis Phaseolus radiatus and Phaseolus aureus were inoculated on to the ovaries of the corresponding plants. Plants grown from seed from the inoculated ovaries produced nodules and grew well.

271.
GUPTA, O.R.D.; JHA, K.K.; SETHI, A.S. and SAHARAN, G.S. (1977) Presence of Azotobacter in relation to physico-chemical properties of some North Indian soils.

Proc. Indian Acad. Sci. 86 B : 389-395.

In cultivated soils including a sample of a "floating island", A. chroococcum and A. agilis were the main species in all samples, including acid ones. Multiple regression indicated that pH and the content of exchangeable Mg were the main factors determining the size of the Azotobacter populations. In aerated N-free culture medium, Mo increased N fixation by Azotobacter.

272.
GUPTA, P.C. (1967) Some additions to Indian soil fungi.

Curr. Sci. 36 : 415.

Fungi isolated from the rhizosphere of Withania sonnifera Linn. included Amerosporium applanatum Berg. et curt., Volutella allii Patters; and V acalyphae Atk., and from the rhizosphere of Datura alba Nees., Helminthosporium Syd.

273.
GUPTA, P.C. (1971) Foliar spray of gibberellic acid and its influence on the rhizosphere and rhizoplane mycoflora.

Pl. Soil 34 : 233-236.

In pot experiments, the variation in the number of fungi/g of dry soil in the rhizosphere was significant in the case of Datura alba Necs. and Ocimum sanctum Linn., and in significant in Whitania sommifera Dun. Qualitatively, no marked difference was observed in the rhizosphere fungal flora of sprayed and control plants.

274.
GUPTA, R.C.; TANDON, R.N.; ARORA, D.K.; SHUKLA, A.N. and SINGH, D.B. (1979) Dynamics of microfungi in fallowland soils with emphasis on volatile fungistasis.

Rev. Ecol. Biol. Sol. 16 : 491-497.

The fungal flora of fallowland soils of Varanasi, India which are highly alkaline and low in organic matter and total nitrogen, is represented by 24 genera and 45 species, dominated by Aspergillus spp. Reduced spore germination of Pestalotia elasticae and Aspergillus humicola demonstrated the volatile fungistatic activity of these soils. Sterilized soil colonized by individual Streptomyces spp. was fungistatic, indicating the role of this genus in fungal inhibition.

275.
GUPTA, R.D. and SAHARAN, G.S. (1974) Microbial ecology in acidic soils of Kangra.

J. Indian Soc. Soil Sci. 22 : 172-175.

No significant correlation was found between soil pH and bacterial count in the eighteen acidic soils of Kangra District, Himachal Pradesh. All the soil samples from different localities showed the presence of spore forming and ammonifying bacterie. Azotobacter were present in all but two highly acidic soils namely Tethal and Sungal. Higher population of actinomycetes was observed in relatively less acidic soils as compared to the more acidic ones. On the contrary fungi were more numerous in strongly acidic soils. In all twenty two microbial species of bacterial, actinomycetes and fungi were identified in these soils.

276.
GUPTA, R.D and SAHARAN, G.S. (1975) Microbiological properties of certain soils of Himachal Pradesh.

Technology, India 12 : 38-41.

Although a large variation in bacterial population was observed (2-24 million/g), with a slightly higher population in neutral soils than in acid soils, no strict correlation existed between soil pH and bacterial count. Liming was essential on acid soils to accelerate the availability of P for microbial activity, mainly of actinomycetes (Streptomyces and Micromonospora), and of Azotobacter.

277.
GUPTA, R.D.; JHA, K.K.; SHARMA, P.K. and SUD, R.D. (1980) Distribution of microorganisms in relation to physico-chemical properties of soils of Jammu and Kashmir. J. Indian Soc. Soil Sci. 28 : 259-262

Soil reaction was found to be negatively related with bacterial population. The higher soil pH favoured the growth of actinomycetes and a lower pH that of fungal population. CEC and Organic carbon content of the soils showed highly significant and positive correlation with the bacterial and fungal population.

278.
GUPTA, R.K. and TALPASAYI, E.R.S. (1984) Nitrate reductase activity of the Cyanobacterium Synechococcus cedrorum.

Z. Allogem. Microbiol. 24 : 283-287

Light incubation invariably stimulated enzyme activity, except in the POLYM-R strain where it remained unaltered. The effect of light was most significant in chlorate-resistant strains. Addition of sulphate to the incubation medium increased the enzyme activity but $NADH_2$ had no effect.

279.
GUPTA, S.R.; RAJVANSHI, R. and SINGH, J.S. (1981) The role of the termite Odontotermes gurdaspurensis (Isoptera: Termitidae) in plant decomposition in a tropical grassland.

Pedobiologia. 22 : 254-261

The role of the termite, Odontotermes gurdaspurensis, in plant decomposition was studied in a tropical grassland by evaluating seasonal activity, food preferences of some "natural" plant materials and the effect of termite activity on physical, chemical and biological properties of soil. Termite activity continued throughout the year with seasonal fluctuations in intensity. Activity was maximum during early winter months (October to November). The observations on food preference showed that among the shoot and root materials of Chenopodium album, Desmostachya bipinnata and mixed-grass samples, the material of Ch. album was consumed very rapidly and that of D. bipinnata was attacked slowly. The soil of earthy galleries present on the soil surface, litter and standing dead shoots contained higher contents of carbon, nitrogen, exchangeable calcium, and total soluble salts compared with that of the adjacent unmodified soil. The total number of microorganisms was significantly higher in termite modified soil than in unmodified soil.

280.
GUPTA, S.R.S. (1956) A new factor in the study of nitrogen

fixation in soil. Indian J. mycol. Res. 2 : 29-36.

Manured alkaline clay soil and sandy acid soil had greater N-fixation than the unmanured soils. Fixation varied with the size of inoculum. With small inoculum fixation was identical for the soils; with large inoculum, fixation increased in manured soil and decreased in the unmanured. Small inoculum fixation indicated the N-fixing capacity of the strain, while large inoculum indicated the N-fixation index of the soil. Non-N fixing organisms and the seasons influenced fixation, the activity was low during the winter months.

281.
GUPTA, V.K. (1974) Effect of foliar application of subamycin on rhizosphere and rhizoplane mycoflora.

Indian Phytopath. 27 : 267.

In a pot experiment, spraying *Trigonella foenumgraecum* seedlings with aqueous solutions of 10 ppm of subamycin on two consecutive days at each of three stages of development (preflowering, flowering and fruiting) increased the number of fungi/g of dry soil, but spraying with 20 ppm decreased it. Spraying affected the species composition of the fungal populations.

282.
GUPTA, V.K. (1974) Effect of seed of *Trigonella foenum graecum* on the growth its *rhizobium.*

Indian Phytopath. 27: 403-404.

Maximum inhibition of *Rhizobium* growing in culture was shown by crushed seed-coats of *T. foenumgraecum,* followed in order by seed minus coat, whole seed, and seed diffusate. Sterilized seeds caused no inhibition.

283.
GUPTA, V.K. (1974) Effect of subamycin on nodulation in *Trigonella foenum-graecum.*

Indian Phytopath. 27 : 262.

In a pot experiment, spraying seedlings with a solution containig 5 ppm subamycin stimulated nodulation, but 10 and 20 ppm solutions were inhibitory.

284.
HEGDE, S.V. and RANGASWAMI, G. (1969) Studies on the thermophilic microorganisms from ten substrata. I - occurrence, isolation and description of thermophilic fungi.

Mysore J. agric Sci. 3 : 198-214.

Among 10 different substrate tested, the numbers of thermophilic fungi were greatest in horse manure, compost and sheep manure. Soils (red and black) had a meagre population. Aspergillus fumigatus, Mucor pusillus and Humicola lanuginosa were present in all cases. Thermoascus aurantiacus was isolated only from soil. Humicola grisea had not previously been reported in these soils. Of the 23 fungi isolated 16 were obligately thermophilic, 3 were facultatively thermophilic and 4 were thermotolerant. The morphological and cultural characters of the fungal isolates were studied and their systematic positions determined. Of the 4 isolating techniques tested, the dilution plate technique with Emerson's modified yeast starch agar, and Cooney and Emerson's technique were the best.

285.
HUTCHINSON, C.M. (1911) The influence of bacteria upon soil fertility. Agric. J. India 6 : 97-113.

A normal soil contains bacteria which may be roughly divided into two classes so far as fertility is concerned (1) beneficial, (2) detrimental. The number of bacteria found depended on factors such as water supply, temperature and organic matter present. The paper suggests two alternative methods for estimating the nitrifying power of a soil. Methods for determining the ammonifying power of soil are also described.

286.
HUTCHINSON, C.M. (1919) Nitrogen fixation in Indian soils. Agric. J. India 14 : 215-219.

Observations made at Pusa of the rate of fixations of N both in field soils and in the laboratory confirmed that not only is the amount of N taken from the air and added to the soil by Azotobacter is of considerable importance, but that this amount varies within wide limits. Algae were found fulfilling their natural function of adding to the organic matter content of the soil by taking the necessary carbon from the CO_2 of the air.

287.
INDIRA, P.U. (1968) Human pathogenic fungi from soil.

Proc. Indian Acad. Sci. 67 : 68-76.

51 soil samples from S. India were examined for pathogenic bacteria especially dermatophyces. Microsporum gypseum was found in nine samples and M canis, Trichophyton sp., Pyrenochaeta sp. and Aspergillus fumigatus were found in four other samples. Ten of the 13 positive samples were collected near human or animal habitats and the other there were from cultivated soil. No pathogenic bacteria were found in forest soils or soils poor in organic matter.

288.
ISWARAN, V. (1958) Effect of humus of legume and non-legume origin on the nitrogen fixation in Azotobacter chroococcum. Curr. Sci. 27 : 489-490.

N fixation in culture solution was increased by the addition of humus produced by the decomposition in soil of sub-aerial residues of dhaincha, sunnhemp and particularly guar, but was not increased by maize or paddy husks.

289.
ISWARAN, V. (1960) Effect of humified organic matter on nitrogen fixation by Azotobacter.

J. Indian Soc. Soil Sci. 8 : 107-113

Humus was extracted from several soils and tank silts and the effect of addition of these substances on nitrogen fixation by addition of these substances on nitrogen fixation by Azotobacter were studied. Humus was also obtained from different plant materials, legumes and non-legumes, after they were allowed to decompose in Delhi soil for a period of two months. Their effects on nitrogen fixation were also studied.

Increase in nitrogen fixation was observed when the humus from soils and tank silts were added in small quantities to the medium inoculated with Azotobacter. In the case of humus from most soils and from all the tank silts, increases in nitrogen fixation were small. The humus from Travancore and Jorhat soils which were acidic and rich in organic matter exerted a marked stimulatory effect on nitrogen fixation.

Humus obtained by incorporating into the soil, leguminous plants like guar, dhaincha and sunnhemp gave greater stimulation of nitrogen fixation than that from non-legumes like maize and paddy. Stimulation of nitrogen fixation than that from non-legumes like maize and paddy. Stimulation of nitrogen fixation could not be correlated with the nitrogen content or the C/N ratio of the humus.

Calcium saturated humus exerted almost the same stimulatory effect on nitrogen fixation by Azotobacter as the humus from natural soil.

290.
ISWARAN, V. (1960) Influence of combined nitrogen on fixation of nitrogen by Azotobacter. Proc. Indian Acad. Sci. 51 : 103-115

Azotobacter was incubated at 32°C for 21 days with 0-10 mg/100 cc of $(NH_4)_2SO_4$, KNO_3, aspartic acid or peptone. The added N inhibited N fixation; the relation between amounts of N added and amounts fixed is expressed mathematically. NH_4-N had the smallest effect and NO_3-N had the largest; there were strong effects with peptone and, at the higher concentrations, with aspartic acid, but small amounts of aspartic acid stimulated fixation.

291.
ISWARAN, V. (1964) Nitrogen fixation by Azotobacter as influenced by soil extracts. Indian J. Appl. Chem. 27(2) : 98-100.

Mo and P had a positive effect on growth and development of Azotobacter in a study made on 21 Indian soils. No correlation was established between N-fixation by Azotobacter and the total soluble salt content, pH, or organic C or N content of the soils.

292.
ISWARAN, V. (1971) Influence of storage temperature of peat on nodulation and growth of pea (Pisum satium).

Mysore J. agric. Sci. 5 : 493-494.

Inoculants stored at 4°C should be incubated for at least 1 week at 26°C before use.

293.
ISWARAN, V. and APTE, R. (1970) Carrier materials for Rhizobium.

Indian Agric. News Digest 2 (4) : 131-132

The results show that charcoal, lignite or FYM could very easily replace peat with addition of energy source and any one of the mixtures could be satisfactorily used as a carrier only for fast growing rhizobia depending on the availability of the materials. Even after 8 weeks, the minimum standard of 100 million rhizobia per g of the inoculant is maintained by these mixtures which appear to be reasonable and practical to bring about nodulation when the seed is inoculated.

294.
ISWARAN, V. and CHHONKAR, P.K. (1971) Note on the comparative efficiency of slurry and sprinkle methods of legume inoculation.

Indian J. agric. Sci. 41 : 1023-1024.

No distinct differences in survival of Rhizobium leguminosarum, Rhizobium sp. (gram group) and R. trifolii on seed coats when inoculated by the slurry method (adhesive solution applied to the seed) or by the sprinkle method (seed dusted with the inoculant after impregnation with 10% Jaggery or 40% gum) were observed.

295.
ISWARAN, V. and RAO, W. (1964) Cobalt in nitrogen fixation by Azotobacter chroococcum. Nature 203 (4944) : 549.

Tests were made with 0.02-5.0 ppm Co in Jensen's medium. At 0.1 ppm, Co enhanced N-fixation by about 50%. N-fixation was slightly less at 1.0-5.0 ppm Co. The effect was not due to traces of Mo present. Mo at 1.0-5.0 ppm also enhanced N-fixation by A. chroococcum.

296.
ISWARAN, V. and ABHISWAR SEN (1958) Effect of salinity on nitrogen fixation by Azotobacter sp. in some Indian soils.

J. Indian Soc. Soil Sci. 6 : 109-113.

Studies were made on salt tolerance of Azotobacter isolated from some soils of different parts of India.

Sodium chloride was found to affect nitrogen fixation by Azotobacter adversely. The concentration of sodium chloride which inhibited fixation of nitrogen completely was different for different strains.

The rate of decrease of nitrogen fixation with increase in the concentration of sodium chloride was less for Azotobacter isolated from soils of higher salinity than for the organisms from less saline ones.

The concentration of sodium chloride which inhibited fixation of nitrogen was higher for Azotobacter isolated from soil containing more soluble salts than for the orgaism from less saline soils.

297.
ISWARAN, V. and SEN, A. (1958) Inactivation of Azotobacter by heat. Curr. Sci. 27 : 341-342.

The N-fixing capacity in culture solution of various strains of Azotobacter chroococcum isolated from different Indian soils was increased by prolonged freezing up to 35 days, though some temporary decrease occurred in most strains exposed to short periods of freezing. Exposure to 45°C for four days and longer decreased N-fixation, which ceased altogether after a fortnight when the organisms took up a coccoid form of diminished size and did not show any raised and pigmented growth. Simulating temperature variations which prevail during summer days in the northern Indian plains by daily incubation for two weeks at 45°C for 5 hours, 32°C for 16 hours and 20°C for 3 hours decreased but did not completely inhibit N fixation by azotobacter.

298.
ISWARAN, V. and SEN, A. (1959) An Azotobacter sp. in the swollen roots of Cyperus rotundus. Curr. Sci. 28 : 449-450.

The organism was isolated in pure culture on mannite agar. Its morphological, cultural and biochemical characteristics are described. From these it is tentatively considered as a strain of Azotobacter sp. through it differs from the common species in its ovate shape of pointed colonies and wide range of pH tolerance.

299.
ISWARAN, V. and SEN, A. (1961) Azotobactor spp. in the rhizosphere of some agricultural crops. Proc. Indian Acad. Sci. 53B : 182-186.

Although no differences in the numbers of azotobacter occurred between rhizosphere and non-rhizosphere soil in wheat and guar fields, the number were greater in the rhizosphere wheat had no effect on N-fixing capacity. Under maize, berseem and guar, azotobacter had a higher N-fixing capacity inside the rhizosphere than outside it. There was no difference in the N-fixing capacity of azotobacter under good and bad maize stands.

300.
ISWARAN, V. and SUNDARA RAO, W.V.B. (1960) The effect of trace elements on nitrogen fixation by Azotobacter. Proc. Indian Acad. Sci. 51B : 103-115.

Traces of Mo, Co, Ni and Fe stimulated N fixation significantly in a medium allowing growth of azotobacter. Mo and Co gave maximum N fixation at 0.02 ppm and Ni and Fe at 1.0 ppm. Combination of Co, Mo, Ni or Fe gave lower N-fixation than Mo alone did. There was evidence of Mo and Co absortption in the cells of azotobacter.

301.
ISWARAN, V. and SUNDARA RAO, W.V.B. (1961) Influence of soil solution on nitrogen fixation by Azotobacter. Indian J. Microbiol. 1 : 27-32.

Extracts obtained with hot water, possibly because of higher P content, showed the greatest stimulation of growth. Analyses of the soil extracts for Fe, P, Mn, total mineral matter, organic C, and loss on ignition are given. Growth of the cultures was depressed as the mineral matter increased, and stimulated by increasing amount of organic C.

302.
ISWARAN, V. and SUNDARA RAO, W.V.B. (1964) Role of cobalt in nitrogen fixation by Azotobacter chroococcum.

Nature 203 : 549.

As little as 0.1 ppm Co increased the amount of N fixed. At 1.0 ppm and at higher concentration the response was less. Mo at concentrations of 1.0 ppm and above accelerated N-fixation.

303.
ISWARAN, V.; APTE, R. and SEN, A. (1973) Use of sodium lactate for culture of Rhizobium spp.

Zentbl. Bakt. ParasitiKde. Abt. II, 128 : 240-242.

Sodium lactate is as good a carbon source for Rhizobium as mannitol. It also maintains the moisture content of peat, enabling Rhizobium to survive at higher temperatures.

304.
ISWARAN, V.; GRAHAM, P.H. and JAUHRI, K.S. (1969) Modified Burton's medium for large-scale production of culture of Rhizobium japonicum.

Curr. Sci. 38 : 346.

Burton's medium had been modified and the modified medium has been found satisfactory for large scale production of cultures of Rhizobium japonicum.

305.
ISWARAN, V.; PATIL, B.D.; JAUHRI, K.S. and SEN, A. (1980) A study of nitrogen fixation by bacteria in some non-leguminous plants. Zentbl Bakt. ParasitKde Abt. II, 135 : 392-395

Nitrogen-fixing bacteria have been isolated from Amorphophallus campanulatus, Colocasia esculenta, Ipomoea batatas, and Monochoria vaginalis . The characteristic of these micro-organisms were studied. It was found that many bacteria were facultative aerobes. The nitrogen-fixing capacity varied from 0.89 to 17.17 mg per g of sugar within a fortnight in Jensen's medium under aerobic conditions, and from 0.21 to 14.7 under anaerbic conditions.

306.
ISWARAN, V.; RAO, A.V. and SEN, A. (1971) Effect of sodium molybdate with inoculant on the yield of pea crop,

Zentbl. Bakt. ParasitKde. Abt. II, 126 : 778-779.

In a pot experiment with Delhi soil (pH 7.2), addition of 14% Na_2MoO_4 to the inoculant (Rhizobium leguminosarum) significantly increased the dry weight of the plants and N uptake.

307.
ISWARAN, V.; RAO, W.V.B.S. and JAUHRI, K.S. (1970) Effect of temperature on survival of Rhizobium japonicum in soil and peat.

Mysore J. agric Sci. 4: 105-107.

When R. japonicum was added soil at 28 or 35°C, counts were appreciable up to the 2nd week, but subsequently decreased suddenly. When added to soil at 40°C rhizobia died rapidly. In peat at 35°C however the survival of rhizobia was high and sufficient to bring about effective nodulation. Use of peat-based cultures for soyabeen sown in June or July is therefore suggested.

308.
ISWARAN, V.; SARMA, K.S.B. and COMHAIRE, M. (1970) Soil fertility, legumes and rhizobium efficiency. I. Introduction. Effect of organic matter and soil reaction.

Agri. Digest No. 19 : 3-19.

Incorporation of legumes in crop rotation has been found to improve soil fertility. The N-fixing efficiency of Rhizobium is improved by addition of organic manures. High pH adversely affected nodulation.

309.
ISWARAN, V.; SEN, A. and APTE, R. (1972) Plant compost as a substitute for peat for legume inoculants.

Curr. Sci. 41 : 299.

After leaves fallen from wayside tree had been composted with 1% of $(NH_4)_2SO_4$, 0.1% superphosphate, 0.01% Na molybdate and water, the compost was dried, ground, mixed with

20% charcoal, 1% $CaCO_3$ and 0.05% K_2HPO_4. The numbers of Rhizobium trifolii, R. leguminosarum and R. japonicum in the mixtue 4,6 and 8 weeks after inoculation were as high as those in inoculated neutralized peat.

310.
ISWARAN, V.; SEN, A. and SINGH, N.P. (1972) Influence of sucrose spraying on nodulation and yield of Phaseolus aureus L.

Pl. Soil 36 : 717-718.

Spraying pot grown Phaseolus aureus plants with 1.0% sucrose solution increased nodulation, grain yield and nitrogen uptake.

311.
ISWARAN, V.; SUBBARAO, N.S. and SUNDARA RAO, W.V.B. (1965) Sodium chloride tolerance by Azotobacter chroococcum.

Curr. Sci. 35 : 126-127.

Limit of tolerance to NaCl concentration by A. chroococcum varied with species between 2.50 and 3.0% NaCl. Significant decrease in N-fixation occurred at concentrations of 0.2% with a Delhi strain and 0.8% NaCl with a Cuttack strain. Cuttack and Samalkot strains retained a capacity to fix 65-80% N under a 1.5% NaCl concentration.

312.
ISWARAN, V.; SUNDARA RAO, W.V.B. and MATHUR, S.P. (1960) Role of vitamin B_{12} in nitrogen fixation by Azotobacter chroococcum. Curr. Sci. 29 : 63-64

When vitamin B_{12} was added to A. chroococcum cultures at concentrations ranging from 0.01 to 0.1 ppm, there were significant increases in N-fixation even without Mo in the medium.

313.
ISWARAN, V.; VIMAL, O.P. and APTE, R. (1971) Pelleting of

soyabean (Glycine max) seeds with activated charcoal and humus.

Zentbl. Bakt. ParasitKde Abt. II, 126 : 45-47.

Pelleting with charcoal, peat humic acids, neutralized peat and untreated peat, each containing as active culture of Rhizobium japonicum, markedly increases number and dry weight of nodules, and green and dry weight of plants.

314.
IYENGAR, M.R.S. and HORA, T.S. (1959). Nitrate oxidation by soil fungi. Naturwissenschaften 46 : 211.

Several isolates of Aspergillus flavus from Delhi soils were capable of forming NO_2 and upto about 45 ppm NO_3 in 16 days in media containing peptone. Where $NaNO_2$ was the sole source of N in the medium, A. flavus produced only 5 ppm NO_3 in 16 days as compared with 50-60 ppm NO_3 by Penicillium sp. in about the same period. An unidentified pycnidium forming fungus resembling Penicillium sp. in its formation of NO_3 from NO_2 and several other fungi were capable of bringing about one or both steps of nitrification.

315.
JADHAV, S.W.; KONDE, B.K. and DAFTARDAR, S.Y. (1973) Amelioration of a saline-sodic soil with Thiobacilli. J. Indian Soc. Soil Sci. 26 : 228-229.

Application of either thiobacilli or sulphur did not affect the soil pH appreciably. However, inoculation of thiobacilli along with sulpher application decreased soil pHs from 9.4 to 7.2 in about 42 days after which the pH changed only slightly. The increases in EC and sulphate content of the soil were observed after the incubation period due to sulphur plus thiobacilli treatments. These changes were possibly due to enhanced microbial activity, which ultimately resulted in decrease in ESP. Thiobacillus novellus was more effective than thiobacillus thiooxidants.

316.
JADHAV, S.W.; KONDE, B.K. and DAFTARDAR, S.Y. (1979) Note on effect of sulphur, *thiobacilli* and cropping on microbial population in a saline-sodic soil. *Curr Agric.* 3 : 115-118.

Addition of sulphur and *thiobacilli* decreased the population of bacteria and actinomycetes by 5-9 and 3-4 times, respectively over control than when they were applied alone; but the population of fungi and N-fixers increased when *thiobacilli* alone were introduced into soil and decreased when sulphur was applied either alone or with *thiobacilli.* While studying the effect of wheat cropping on microflora, a decrease in the population of actinomycetes, fungi and N-fixers was observed to a considerable extent due to the influence of sulphur alone and *thiobacilli* alone treatments.

317.
JAGTAP, B.K. and SHINGTE, A.K. (1982) Influence of *Azotobacter* inoculation in conjunction with farmyard manure in economising use of fertilizer nitrogen to wheat. *J. Indian Soc. Soil Sci.* 30 : 210-212.

Pooled results of analysis indicate a significant response of wheat to *Azotobacter* inoculation alone without FYM application. Result show that with the inclusion of FYM even at the lower dose of 5 t/ha and bacterization, there could be a saving of fertilizer to the extent of 60 kg N/ha.

318.
JAIN, J.M., NARAYANASAMY, G., SARKAR, M.C. and DATTA, M.N. (1980) An evaluation of nitrification retardation property of *Citrullus colosynthis* cake and its influence on yield and N uptake by wheat.

J. Indian Soc. Soil. Sci. 28 : 480-484.

Citrullus colosynthis oil cake was tested as nitrification retarder under laboratory and green house conditions. The same showed NH_4-N conservation when applied at the rate of 5 and 10% by weight of urea. Higher rate (20%) of the cake resulted in NO_2-N accumulation leading to lower N recover by wheat (var. Kalyansona). The cake also proved a good source of nitrogen and was nearly 67 per cent as efficient as urea.

319.
JAIN, M.K. and REWARI, R.B. (1974) Isolation of seed-borne microflora from leguminous crops and their antagonistic effect on *Rhizobium.*

Curr. Sci. 43 : 151.

In vitro, seeds of mung (*Phaseolus aureus*), urd (*Phaseolus mungo*) and soyabean were found to harbour different types of bacteria, fungi and actinomycetes. The number of isolates varied from crop to crop and from one variety to another in the same crop. Some bacterial and actinomycete isolates from all three crops showed antagonism to their respective *Rhizobium.* However, no antagonistic fungal isolate was found from soyabean and urd, and only one from mung.

320.
JAIN, M.K. and REWARI, R.B. (1976) Studies on seed coat toxicity to rhizobia of urid (*Phaseolus mungo*), mung (*Phaseolus aureus*) and soyabean (*Glycine max.*).

Zentble. Bakt. ParasiKde Abt. II, 131 : 163-169.

Seeds or urid, mung, and soyabean inhibited their specific *Rhizobium* when the latter come into contact with the seed surface. This seed coat toxicity which not only varied from crop to crop but also from variety to variety, was attributed to the diffusable toxic substances of the seed coat. Presoaking of the seeds in water for varying periods of time mitigated seed coat toxicity appreciably. Presoaking was also found to have considerable influence on the survival of the Rhizobium in the seed and the total number of nodules produced, although in many cases the total yield did not reflect the effect of presoaking, possibly because of climatic and edaphic factors.

321.
JAIN, M.K. KAPOOR, K.K.; and MISHRA, M.M. (1979) Cellulase activity, degradation of cellulose and lignin, and humus formation by thermophilic fungi.

Trans. Brit. Mycol. Soc. 73 : 85-89

Several species of thermophilic fungi isolated from soil, compost heaps and decomposing organic debris exhibited cellulase activity when grown on wheat straw, but only *Humicola insolens*-1 and *Aspergillus fumigatus* exhibited significant cellulase activity when grown on cellulose powder. *Humicola insolens*-1 and 2 and *Torula thermophila* degraded cellulose and lignin and reduced the weight of the substrate. *H. lanuginosa*, *Mucor miehei* and *Talaromyces dupontii* degraded lignin but not cellulose, whereas *A. fumigatus* and *Malbranchea pulchella* degraded cellulose more than lignin. Humus formation was greatest by *A. fumigatus* and *H. insolens*-1.

322.
JAISWAL, S.P. (1967) Effect of some pesticides on soil microflora and their activity.

J. Res. PAU 4 : 223-226.

The gamma isomer of benzene hexachloride, heptachlor and telodrin, even at levels much higher than those recommended, did not adversely affect the microbial population and appeared to stimulate N-fixation, nitrification and P-mineralization.

323.
JAUHRI, K.S. and AGARWAL, D.K. (1984) Effect of seed dressing fungicides on soyabean inoculated with *rhizobium japonicum* in charcoal rot infested soil.

Pesticides 18 : 47-49

Four seed dressing fungicides namely Thiram, Captan, PCNB and Dithane M-45 were tested for their relative efficacy against the charcoal rot (*Macrophomina phaseolina*) on soyabean with special reference to seeds treated with rhizobial inoculants. Results showed that Dithane M-45 alone or in combination with Thiram gave the best yield and seed emergence in M. Phaselina infested soil when the seeds were also inoculated with R. japonicum. No adverse effect of fungicides on inoculation was observed.

324.
JAUHRI, K.S. and PHILIP, K. (1983) Major constraints and remedies in the production and storage of bacterial inoculants in India.

Sci. Cult. 49 : 403-404.

The efficiency of presumed as a carrier of bacterial inoculants has been found to improve by mixing it with finely powdered charcoal (75 : 25). An earthern incubator has also been devised for catering to the needs of rural areas where no refrigeration facilities are available.

325.
JAUHRI, K.S. and PHILIP, K. (1984) Presmued : a potential carrier for Rhizobium and Azotobacter. II - Influence of storage temperature on the survival of Rhizobium and Azotobacter in different carriers and nodulation efficiency of rhizobial inoculants.

Zentb. Microbial. 139 : 97-107

The nodulation efficiency of various carrier base inoculants, stored under different temperatures, was evaluated in a pot trial in Gycine max Var. Bragg and Vigna radiata var. PS-16. The maximum number of rhizobia on seed and nodules per plant were obtained when inoculants were prepared from pressmud plus charcoal (75 : 25) and stored at 32°C.

326.
JAUHRI, K.S.; BHATNAGAR, R.S. and ISWARAN, V. (1979) Associative effect of inoculation of different strains of Azotobacter and homologus Rhizobium on the yield of moong (Vigna radiata).

soyabean (Glycine max) and pea (Pisum sativum).

Pl. Soil. 53 : 105-108

In V. radiata and soyabean, use of a suitable strain of Azotobacter gave higher yields than the use of Rhizobium as inoculant. In V. radiata even a strain of Azotobacter isolated from the rhizosphere of berseem gave similar yields to Rhizobium. In pea, association of Rhizobium with a strain of A. chroococcum isolated from the rhizosphere of pea gave higher yield of Rhizobium alone. It may be possible that higher yield could be obtained when a suitable strain of Azotobacter obtained from screening a large number of strains of Azotobacter from the rhizosphere of pea is used.

327.
JAUHRI, K.S.; PHILIP, K. and SUBBA RAO, N.S. (1983) An inexpensive device for storing inoculants in tropical countries.

Agric. Sci. Digest. 3 : 175-176.

In tropical countries, nonavailability of cheap cold storage facility at farmers level is a major constraint in proper storage and commercialization of bacterial inoculant. Visualizing the need of rural masses, a cheap earthern inoculator has been devised by IARI, New Delhi. The device maintains a favourable temperature for storing inoculants.

328.
JAYACHANDRAN, S. and CHANDRAMOHAN, N. (1977) Influence of insecticides on the phyllosphere and rhizosphere of rice crop.

Il Riso 26 : 323-330.

Three contact insecticides (diazinon, chlorfenvinphos and quinalphos) and four systemic insecticides (AC 92100, mephosfolan, carbofuran and chlorodimeform) were applied at 1 kg. a.i./ha on the 20th and 50th day after transplanting. Phyllosphere and rhizosphere samples were collected on the fifth day after each application. Both the contact and the systemic insecticides increased the number of micro-organisms in the rhizosphere at the first sampling, but at the second sampling only the contact insecticides did so. There was no correlation between the rhizosphere population and soil enzyme activity.

329.
JAYACHANDRAN, S.; BALASUBRAMANIAN, A. and PURUSHOTHAMAN, D. (1978) Influence of some phenotic compounds on the growth, polysaccharided production and nitrogen fixation by Azotobacter chroococcum Beij. pp.248-252. In S.P. Sen, Y.P. Abrol and S.K. Sinha (eds.), Nitrogen Assimilation and Crop Productivity. Associated Publishing Company, New Delhi.

Of the various phenolic compounds tested hydroquinone, 8-hydroxy quinoline, protoatechuic acid and pyrogallol stimulated the growth of A. chroococcum while querecetin, 1, 5-dihydroxy anthroquinone and phthallic acid were inhibitory. There was no correlation between growth and polysaccharide production.

330.
JAYARAMAN, K.N. and PRASAD, N.N. (1972) Production of phosphatase by soil Aspergilli.

Madras agric. J. 59 : 640-641.

In fifteen cultures of Aspergillis spp. isolated from field soil, A. niger strains showed maximum phosphatase activity.

331.
JHA, K.K. (1969) Effect of vanadium and tungsten on nitrogen fixation and the growth of Medicago sativa. J. Indian Soc. Soil Sci. 17 : 11-13.

In sand culture, it was found that vanadium had no stimulatory effect on nitrogen or the growth of Medicago sativa inoculated with an effective strain of Rhizobium meliloti or supplied with ammonium nitrate. At the level of 500 ppm it reduced the plant growth, the inhibitory effect being particularly severe on the root. On the other hand tungsten increased nitrogen fixation and the dry matter yield of the inoculated plants. The results are suggestive of a direct role of tungsten in symbiotic nitrogen fixation.

332.
JHA, K.K., ALI, M.A., SINGH, R. and BHATTACHARYA, P.B. (1965) Increasing rice production through the inoculation of Tolypothrix tenuis, a nitrogen fixing blue-green alga.

J. Indian Soc. Soil Sci. 13 : 161-166

Pot and field trials conducted for 7 years in a neutral loamy soil have shown that it is possible to increase the yield of rice through soil inoculation with Tolypothrix tenuis - a blue-green alga. This effect may be due to the substantial amounts of nitrogen fixation and the addition of organic matter by this organism. It is also likely that a good portion of the nitrogen fixed by this alga is excreted into the soil.

333.
JOHRI, K. (1981) Interactions between soil Fusaria and thermophilic fungi.

Curr. Sci. 50 : 966-967

Five species of Fusarium were recorded from cultivated soil (wheat-linseed rotation). The inhibition by thermophilic micoflora in arable soil appear to be of little consequence.

334.
JOHRI, K.; JOHRI, B.N. and SAKSENA, S.B. (1975) Colonization capacity of the soil micro-organisms in relation to fungistasis.

Pl. Soil. 43 : 347-354.

Aspergillus fumigatus was the pioneer colonizer, followed in the order of decreasing colonizing capacity by, A. flavus, A. niger, A. terreus and Penicillium nigricans. Sensitivity of test spores to soil fungistasis declined in the order, C. geniculata - Pestalotia sp - A. tenius - H. rostratum. Studies conducted with an inoculum of four soil fungi (A. fumigatus, A. flavus, A. niger, and P. nigricans) as well as those involving fumigation of an Evans tube by CS_2 also proved the greater colonizing capacity of A. fumigatus. The level of fungistasis closely followed the colonization pattern of this fungus. When added individually to the soil, all fungi were able to inhibit spore germination of test fungi to some degree.

335.
JOHRI, K.; SAKSENA, S.B. and JOHRI, B.N. (1975) Soil mycoflora as producer of volatile fungistatic factors.

Proc. Indian natn. Sci. Acad. B 41 : 148-154.

Ten soil samples collected from Madhya Pradesh and South India were investigated for their fungistatic properties. Spore germination of Alternaria alternata, Curvularia geniculata, Drechslera rostrata and Pestalotia sp. was effectively checked over these soils when tested for fungistasis by the cellophane agar-disc and U-tube techniques. The mycoflora of the soil samples was studied and Aspergillus spp. were found to be predominant. The production of volatile fungistatic factors by the four commonest species of Aspergillus and four of Penicillium, growing in sterilized soil, was compared. A.fumigatus and P. jenseni caused most inhibition.

336.
JOSHI, I.J. and CHAUHAN, R.K.S. (1981) Soil fungal ecology of cultivated areas of Chambal ravines.

Proc. Indian natn. Sci. Acad. 47 : 248-254

The ecology of soil microfungi was studied in three terraced areas under Triticum vulgare (A), Brassica campestris (B) and Cajanus cajan (C) respectively.

54 fungal species were isolated of which 24 were common to the three fields; 8 species were restricted to A, 7 to B and 4 to C. The fungal population was highest in C and lowest in A. Aspergillus terreus was the most widely distributed species.

Except in the case of calcium carbonate content, edaphic factors showed no significant effect on fungal numbers.

337.
JOSHI, I.J. and CHAUHAN, R.K.S. (1982) Investigations into the soil mycoecology of Chambal ravines of India. I - Fungal communities and seasonal succession.

Pl. Soil. 66 : 329-339.

The seasonal succession of fungal communities in wheat fields in India was investigated. Of the 60 species isolated, 25 showed a wide ecological amplitude while some species were specific to a particular season. The greatest number of species was isolated during the rainy season, and the smallest number during the summer. Aspergillus species were dominant. The effect of climatic conditions on the composition of the fungal community are discussed.

338.
JOSHI, N.V. and BISWAS, S.C. (1938) Cellulose decomposition by a new organism growing is association with other organisms commonly occurring in soils and manures. Proc. 25th Indian Sci. Cong. 226.

The organism decomposes cellulose only when growing in association with other organisms. Its morphological, cultural and biochemical characteristics are described.

339.
JOSHI, N.V. and BISWAS, S.C. (1948) Does photo-nitrification occur in the soil. Indian J. agric. Sci. 18 : 115-129.

A series of carefully controlled experiments gave no evidence of the promotion of ammonification or nitrification by solar radiation but suggested that reduction of NO_3 to NO_2 forms might occur by this means.

340.
JOSHI, N.V. and JOSHI, S.G. (1953) Soil fertility and microbial

activity.

J. Indian Soc. Soil Sci. 1 : 15-20

The fertility levels of the soils as well as crop yields were found to be directly correlated with the microbial activity of the soil. However, no definite correlation was observed between CO_2 evolution and soil productivity. Production of CO_2 in the soil by micro-organisms may lead to a greater solubilization and availability of plant nutrients. The study revealed that a higher rate of nitrification alone need not signify higher productivity, since a renewal of nitrate supply at a slow rate may equally be found to be beneficial to the plant, depending upon the requirements of the plant at different periods in its life process.

341.
JOSHI, N.V. and JOSHI, S.G. (1953) Studies in carbon dioxide evolution in soils. The depression in the production of carbon-dioxide in the soils observed after the addition of mineral fertilizer.

J. Indian Soc. Soil Sci. 1 : 115-119

An experiment was conducted to find out whether CO_2 evolution and crop yields from some soils could be increased to the level obtained in other fertile soils by separately adding the essential nutrients like N and P. The results showed a depression in CO_2 evolution due to addition of mineral salts, mainly because of limited quantity of water in the soil. This probably results in a very high concentration of the salts in the water films surrounding the soil particles in which the organisms are living. The mineral fertilizers get dissolved to the extent that the high concentration of soluble salts in soil solution brought a physiological dryness for some of the micro-organisms, which may collapse and die. This finally resulted in lower production of CO_2 after addition of mineral salts. This high concentration of salts did not caused succumbing of plants to the extent as the microorganisms as plants were irrigated at intervals which enabled the plants to survive and utilize the mineral nutrients afterwards, and make them grow well. This led to absence of correlation between CO_2 production and crop yields. It is also possible that the collodial particles got coagulated due to addition of mineral salts, leading, thereby to decreased surface area available for the activity of micro-organisms, but the adequacy of nutrients helped the plants to grow better.

342.
JYOTISHMATI, U. and BAHADUR, K. (1978) Effect of uranyl and molybdata ions on microbial fixation of nitrogen.

Zentbl. Bakt. ParasitKde Abt. II. 138 : 157-162.

N-fixation by three strains of Azotobacter in liquid culture was increased by the addition of uranyl acetate plus sodium molybdate.

343.
KABI, M.C., POI, S.C. and BHADURI, P.N. (1982) Potential of nitrogen nutrition of leguminous crops through Rhizobial inoculation West Bengal soils.

Trans. 12th int. Congr. Soil. Sci., New Delhi 6 : 52.

The paper deals with a survey of nodulation behaviour in different leguminous crops in West Bengal soils. Extensive studies on this aspect for the last several years indicated the impacts of legume inoculation with Rhizobia on nitrogen economy of the soils.

344.
KADAM, S.S., KACHHAVE, K.G. ; CHAVAN, J.K. and SALUNKHE, D.K. (1978) Effect of nitrogen, Rhizobium inoculation and simazine on yield and quality of bengal gram (Cicer arietimum L.) pp 301 In S.P. Sen, Y.P. Abrol and S.K. Sinha (eds.), Nitrogen Assimilation and Crop Productivity. Associated Publishing Company, New Delhi.

A field experiment was conducted to study the effect of Rhizobium inoculation, nitrogen and simazine, individually and in combination, on yield and quality of Bengal Gram. Application of nitrogen, simazine and seed inoculation with Rhizobium increased the grain yield significantly. The combined treatment of Rhizobium, simazine and nitrogen increased the grain yield to the extent of 70 per cent over control. Application of simazine increased the methionine content.

345.
KALYAN SINGH, RAO, K.L. and SRIVASTAVA, V.K. (1985) Effect of phosphorus and bio-fertilizers on growth analysis of pea (Pisum sativium L.).

Proc. Natn. Sem. Plant Physiol. (Abst.) p. 109

Dry matter accumulation was significantly influenced by phosphorus as well as by bacterial cultures. Relative leaf growth rate decreased with advancement in the age of the crop.

346.
KAMAL and BHARGAVA, K.S. (1970) Studies on soil fungi from teak forests of Gorakhpur. III. Additions to our knowledge of Indian Aspergilli.

Indian Phytopath. 22 : 376-381.

Of the 10 species isolated, 4 species (A. avenaceus, A. subolivaceus, A. terreus var. africanus, A. terreus var. aureus) are reported for the first time in India.

347.
KAMAL and BHARGAVA, K.S. (1970) Studies on soil fungi from teak forests of Gorakhpur. V. A contribution to Indian Torulaceae.

Indian Phytopath. 23 : 558-561.

Descriptions of Torula terrestris, T. herbarum, T. herbarum f. quaternella, T. herbarum f. quaternella (a new strain) and Bahusandhika calgans are presented. The last 3 are new records for India.

348.
KAMAL and GUPTA, M.L. (1974) Studies on Aspergilli from soils of Gorakhpur. III. Some Aspergilli new to India.

Indian Phytopath. 27 : 29-32.

The paper describes four species of aspergilli viz. A. junus var. brevis, A. fischeri var. glaber, A. fischeri var. spinusus and A. itaconicus from different soils of Gorakhpur.

349.
KAMAL and SRIVASTAVA, K.S. (1974) Rhizosphere microfungi of some weeds.

Technology, India 11 : 440-442.

Rhizosphere fungal populations of ten weed plants (five species of Gramineae and five of Cyperaceae) were studied. Most of the fungi isolated were common to the rhizosphere of both plant families, but two Penicillium species and one Fusarium species were restricted to graminaceous rhizosphere and a large number of species (Aspergillus spp., Chaetomium sp. (green), Penicilium sp., Myrothecium sp. and Acrophialophora fusispora) were confined to the rhizospheres of cyperaceous plants.

350.
KAMAL and VERMA, A.K. (1976) Micro-fungal flora in the root region of arhar (Cajanus cajan (l.) Mill).

Fertil. Technol. 13 : 155-157.

The fungal flora of the rhizosphere and rhizoplane of arhar and of non-rhizosphere soil is reported.

351.
KAMAL and VERMA, A.K. (1976) Microfungi from the root region of urd (Vigna mungo (Roxb.) Linn).

Fertil. Technol. 13 : 311-312.

39 forms of microfungi were isolated by soil dilution and plate counts from non-rhizosphere, rhizosphere and rhizoplane regions of the soil associated with urd. The distribution of the species in the three zones is discussed.

352.
KANAKAMBARAN, P.N. and RAJ, J.S. (1974) Effects of collar rot and ring-barking on the rhizosphere microflora and certain chemical constitutents of Sword bean plants.

Agric. Res. J. Kerala 12 : 56-63.

An increase in the microbial population in the rhizosphere of sword bean plants was observed as the plants became older. The increase was more pronounced in plants inoculated with Rhizoctonia sp. than in plants that were ring-barked. There was a marked reduction in the carbohydrate and nitrogen contents in the root system of ring-barked and inoculated plants, and especially in the former. In the inoculated plants the downward movement of food materials was stopped gradually, whereas in the ring-barked plants it occurred abruptly with the result that the food materials present in the root system may have been depleted faster during the normal metabolic activities of the root.

353.
KANAUJIA, R.S. (1977) Studies on certain ecological aspects of soil fungi. VII - Fungi of amended soils.

Fertil. Technol. 14 : 136-139.

Single plots were treated separately with urea, ammonium sulphate, superphosphate, organic manure and fresh twigs and leaves of Ipomoea fistulosa. An additional plot was irrigated weekly. The fungal species and population of the plots are reported.

354.
KANAUJIA, R.S. and MISHRA, R.R. (1977) Observations on soil fungistasis. I - Survey of fungistasis in U.P. soils.

Acta. Bot. Indica 5 : 169-173.

Twenty-one soil samples, collected from eight districts of Uttar Pradesh were assayed for their capacity to bring about fungistasis. The soils exhibited varying levels of fungistasis against two test fungi, viz., Curvularia lunata and Fusarium nivale. The fungal population of the soil samples and their physico-chemical properties were also determined and a correlation between them and soil fungistasis was established.

355.
KANDASAMY, D. and PRASAD, N.N. (1975) Studies on certain factors influencing legume-rhizobia interrelationship. III - Antagonistic soil streptomycetes in the rhizosphere region.

Hindustan Antibiot. Bull. 17 : 115-117

The study indicate the possible scope of foliar application of growth regulators as a means to encourage the occurrence of anti-rhizobial actinomycetes in the rhizosphere.

356.
KANDASAMY, D. and PRASAD, N.N. (1979) Colonization by rhizobia of the seed and roots of legumes in relation to exudation of phenolics.

Soil Biol. Biochem. 11 : 73-75.

A study of the colonization pattern of Rhizobium inoculated either in the soil or on to the seeds of green gram (Phaseolus aureus), black gram (P. mungo) and sun hemp (Crotalaria juncea) revealed a significant reduction in Rhizobium in the spermosphere of sun hemp which was attributed to excretion of phenolic compounds by the seed. As the plants aged, the quantity of phenolics exuded by the roots were reduced and there was a parallel increase in the Rhizobium population in the rhizosphere of all three species.

357.
KANDASAMY, D. and RANGASWAMI, G. (1967) Changes in the

rhizosphere microflora of sorghum due to foliar nutrient sprays.

Indian J. agric. Sci. 37 : 143-150.

Foliar spraying with 0.35 M $(NH_4)_2SO_4$ + 0.455M urea + 0.324M Na_2HPO_4 + 0.256M KCl on 15-75-day-old sorghum strains grown in sandy clay loam generally suppressed the bacterial population in the rhizosphere of Co. 4, but had little effect with K.1 strains. In both strains, fungal populations in rhizospheres of young plants increased. Actinomycete populations sometimes increased considerably. Increased numbers of *Penicillium* in the treated Co. 4 plants and more antagonistic actinomycetes in Co. 4 and K.1 plants treated with urea and KCl were found.

358.
KANDASAMY, D.; MARIMUTHU, T.; OBLISAMI, G. and SUBRAMANIAM, T.R. (1977) Effect of application of insecticides on the HCN content and rhizosphere microflora of sorghum plants.

Madras agric. J. 64 : 302-306

Sorghum plants grown in the field were treated when one week old with carbofuran and phorate granules applied to the soil. The contents of HCN in roots and shoots were determined at 1-20 days after treatment. There was no relationship between root HCN content and populations of fungi, bacteria and actinomycetes in the rhizosphere, except in the case of *Azotobacter* which showed a negative relationship. Monitoring of some fungal genera in the rhizosphere indicated that *Aspergillus* increased when root HCN levels were low.

359.
KANDASAMY, D.; MARIMUTHU, T.; RAJUKKANNU, K.; RAGHURAJ, R.; OBLISAMI, G.; KRISHNAMOORTHI, K.K. and SUBRAMANIAM, T.R. (1975) A study on the relationship between the dissipation of insecticides and rhizosphere microflora of paddy.

Madras agric. J. 62 : 203-207.

Application of granuler phorate to a paddy crop significantly increased the populations of fungi and bacteria in the rhizosphere, and there was a clear relation between the levels of phorate residues and the populations of actinomycetes and Azotobacter. Carbofuran, however, did not show this effect.

360.
KANDASAMY, R. and PRASAD, N.N. (1971) Lignite as a carrier of Rhizobia.

Curr. Sci. 40 : 496.

Rhizobium sp. multiplied well in lignite and the higher count of 13.4 x 10^8 could be increased to 19.15 x 10^8 by amending it with 1% soyabean powder. Inoculated groundnut seeds showed increased growth, total N and number of nodules.

361.
KANNAIYAN, S. and PRASAD, N.N. (1974) Influence of foliar nutrient sprays on the population of Fusarium oxysporum f. melonis and other soil microflora in the rhizosphere of muskmelon.

Indian Phytopath. 27 : 527-531.

Urea, disodium phosphate and potassium chloride were applied individually as foliar sprays at 0.5% concentration to musk melon plants inoculated with F. oxysporum f. melonis. Nutrient sprays increased the fungal, bacterial and actinomycete population. Potassium spray reduced the occurrence of the wilt pathogen and increased total fungi in the rhizosphere.

362.
KANNAIYAN, S. and RAJESWARI, N. (1983) Comparative effect of fertilizer nitrogen and Azolla biofertilizer on tiller production.

Sci. Cult. 49 : 245-246.

Azolla inoculation was effective in increasing the tiller number and provided positive response of the rice crop.

363.
KANNAIYAN, S.; THANGARAJU, M. and OBLISAMI, G. (1984) Influence of inoculum potential on the growth of Azolla.

Sci. Cult. 50 : 228

Results show that the growth rate of Azolla was poor at lower inoculum level. Increase in inoculum level correspondingly increased the growth of Azolla. The threshold level of Azolla inoculum for better growth and multiplication was 200 g/m2 under field conditions. The fresh Azolla inoculum required per hectare of rice field is only 2 t and this has got agronomical significance.

364.
KANNAIYAN, S.; THANGARAJU, M. and OBLISAMI, G. (1984) Influence of neem cake on the growth of Azolla pinnate.

Madras agric. J. 71 : 66-67

Results show that neem cake application not only stimulated the growth of Azolla but also reduced Pyralis sp. infestation.

365.
KANNAIYAN, S.; THANGARAJU, M. and OBLISAMI, G. (1984) Inoculum concentration and growth rate of Azolla.

Madras agric. J. 71 : 64-65

The growth rate of Azolla was poor at lower inoculum levels. Increases in inoculum level correspondingly increased the growth of Azolla. The threshold level of Azolla inoculum could be fixed at 200 g/m2 under field conditions.

366.
KANNAIYAN, S.; VISWANATHAN, G. and RAJESWARI, N. (1983) Influence of green and brown colour variation on NPK contents of Azolla pinnata L.

Sci. Cult. 49 : 251.

The gradual change of green colour to brown colour in Azolla has been found to be due to the accumulation of anthocyanin pigments. There was not much variation in NPK contents of green and brown Azolla. The brown colour change is only a temporary phenomenon in Azolla under stress condition and the plants recover from brown to green colour under favourable environmental conditions.

367.
KAPOOR, K. and SHARMA, V.K. (1980) Effect of certain herbicides on survival, growth, and nitrogen fixation of blue-green algae Anabaena doliolum Bharadwaja.

Z. Allogem. Microbial. 20 : 465-469.

Growth and fixation of nitrogen were retarded in Anabaena doliolum when grown on elemental nitrogen in presence of four different herbicides. The alga was found to be intolerant of these chemicals under nitrogen-fixing conditions even at low concentrations. The herbicides inhibited photosynthesis and phosphorylation processes and probably also nitrogenase activity.

368.
KARANTH, N.G.K. and VASANTHARAJAN, V.N. (1973) Persistence and effect of Dexon on soil respiration.

Soil Biol. Biochem. 5 : 679-684.

The fungicide (p-dimethylaminobenzenediazo sodium sulphonate) retarded the breakdown of glucose and paddy straw added to soil. During the early stages of glucose decomposition and throughout the decomposition period (60 days) of the straw, CO_2 production was inhibited. The fungicide also reduced dehydrogenase activity. After sixty days Dexon had almost disappeared from the soil.

369.
KARANTH, N.G.K. and VASANTHARAJAN, V.N. (1974) Phytotexicity of Dexon (p-dimethylaminobenzenediazo sodium sulfo-

nate) towards the legume Crotolaria juncea.

Proc. Indian natn. Sci. Acad. 40B: 576-585.

Dexon (p-dimethylaminobenzenediazo sodium sulfonate), a soil fungicide used to control 'Damping-off' of seedlings caused by Pythium sp., is applied either as a seed or soil dressing. In pot culture studies Dexon was found to be toxic to the germination of Sunnhemp (Crotalaria juncea) a green manure plant. It was also observed to affect nodulation. Interestingly the growth of Rhizobium in culture was not affected by Dexon. In fact, four strains of rhizobia capable of nodulating Sunnhemp, Soyabean, Bengal gram and Sesbania respectively were resistant to Dixon. However, the fungicide was observed to be toxic to Sunnhemp, even at 100 ppm; particularly the tap root elongation was affected. The influence of Dexon on germinating seeds was therefore studied in detail. The fungicide indeed inhibited the tap root elongation and the inhibition was irreversible by external addition of either 3-indole acetic acid or gibberellic acid at the concentration tried. On the other hand at certain other concentrations Dexon seemed to be able to overcome the toxicity of excess, externally added 3-indole acetic acid. The conclusion that the inhibitory effect of Dexon on nodulation and nitrogen fixation is due to the effect on plant metabolism particularly hormone rather than on rhizobia seems to be plausible from the experimental results so far recorded.

370.
KASI VISWANATH, G. and PILLAI, S.C. (1977) Influence of superphosphate on soil protozoa.

J. Indian Inst. Sci. 59 C : 113-120.

Protozoa, especially species of Colpidium and Colpoda, were very numerous in soil to which superphosphate at 320 kg/ha P_2O_5 had been applied, the total number reaching 195, 000/g of soil.

371.
KASIRAGAN, G.; PURUSHOTHAMAN, D.; SANKARAN, N. and SUBRAMANIAN, V. (1978) Usefulness of Azotobacter in economising nitrogenous fertilizers in crop production. pp. 253-259. In S.P. Sen, Y.P. Abrol and S.K. Sinha (eds.), Nitrogen Assimilation and Crop Productivity. Associated Publishing Company, New Delhi.

The extent to which the free living, N-Fixing, heterotroph, Azotobacter could replace nitrogenous fertiliser for rice crop was investigated. An efficient isolate, AB-3, fixing as much as 18.60 mg of N/g of mannite. Field trial at Agricultural Research Station, Kaverjpattinam recorded a general increase in grain and straw yield due to Azotobacter inoculation. In a field trial at the Tamil Nadu Agricultural University, Coimbatore in which organic matter in the form of paddy straw was incorporated at 5 tons/ha. Azotobacter inoculation strikingly increased the yield and thereby saved the nitrogenous fertiliser to an extent of 25% of the recommended dose. This saving roughly amounts to Rs. 150/ha.

372.
KATHAVATE, Y.V.; SHIVAMURTHY, S.C. and SEN, A. (1964) Influence of molybdenum on some micro-biological properties of soils.

Indian J. agric. Sci. 34 : 245-250.

Up to a concentration of 100 ppm, Mo did not adversely affect the ammonification of organic N. It enhanced the formation of NH_4-N from peptone, caused nitrites to accumulate from NH_4-N and, at 100 ppm, decreased the formation of nitrates from nitrites. N fixation in Delhi soils was not affected by the form or concentration of Mo up to 100 ppm, but was higher in presence of $CaCO_3$.

373.
KAUSHIK, B.D.; BOSE, P. and VENKATARAMAN, G.S. (1974) Isolation and characterization of UV induced mutants of Rhizobia. Proc. Indian natn. Sci. Acad. 40B : 486-487.

The radioresistant mutants of both species of Rhizobia (Trifolii and leguminosarum) were stable and showed increased tolerance to UV irradiation as compared to the wild strains, although the degree of radioresistance varied with the mutant strains. No specific correlation between drug resistance and presence or absence of nodulating capacity could be observed.

374.
KAUSHIK, B.D.; KRISHNAMURTI, G.S.R. and VENKATARAMAN, G.S. (1981) Influence of blue-green algae on saline-alkali soil. Sci. Cult., 47 : 169-170.

Application of blue-green algae @ 15 kg/ha followed by leaching decreased pH2 from 9.4 to 8.0, EC from 23.8 to 9.5 mmhos/cm, ESP from 6.4 to 0.8 me/100 g and increased permeability from 0.7 to 91.8 cm2 x 10-4 and nitrogen from 0.06 to 0.07%. It was more effective than gypsum treatment wherein permeability increased to only 6.9 cm2 x 10-4.

375.
KAVIMANDAN, S.K.; LAKSHMI KUMARI, M. and SUBBA RAO, N.S. (1978) Non-symbiotic nitrogen fixing bacteria in the rhizosphere of wheat, maize and sorghum.

Proc. Indian Acad. Sci. B 87 : 299-302.

In general, the numbers of Clostridium-like anaerobic bacteria were greater than those of Azotobacter chroococcum in the wheat rhizosphere, but in the rhizospheres of maize and sorghum the reverse was true.

376.
KAVIMANDAN, S.K.; SUBBA RAO, N.S. and MOHRIR, A.V. (1978) Isolation of Spirillum lipoferum from the stems of wheat and nitrogen fixation in enrichment cultures.

Curr. Sci. 47 : 96-98.

Enrichment cultures derived from 9 Indian wheat varieties fixed 5.7-22.54 mg. N_2/50 ml of Na malate medium, average fixation being 11.10 mg/50 ml.

377.
KHAN, M.W.; KHAN, A.M. and SAXENA, S.K. (1979) Suppression of phytophagus nematodes and certain fungi in the rhizosphere of okra due to oil-cake amendments. Acta Bot. Indica 7 : 51-56.

Application of oil-cakes of Azadirachta indica, Madhuca indica, Arachis hypogea and Ricinus communis at the rate of 100 lb. nitrogen/acre significantly suppressed the populations of Tylenchorhynchus brassicae, Hoplolaimus indicus, Helicotylenchus erythrinae and Meloidogyne incognita in the rhizosphere of Abeimoschus esculentus in both pot and field experiments. Populations significantly increased in unamended soil or soil amended with ammonium sulphate.

378.
KHANDELWAL, K.C. ; SINGH, D.P. ; KAPOOR, K.K. and MISHRA, M.M. (1978) Nitrification of urea coated with alcohol extract of neem (Margosa) seed cake in soil. pp. 303-304. In S.P. Sen, Y.P. Abrol and S.K. Sinha (eds.), Nitrogen Assimilation and Crop Productivity. Associated Publishing Company, New Delhi.

Hot-alcohol extract of the neem (margosa) seed cake was prepared and coated on urea granules at concentration of 2,4 and 8 per cent (w/w). The Population of nitrifying bacteria and changes in mineral nitrogen were determined in soil incubation experiment. The inhibitory effect of the extract was on the population Nitrosomonas and not on Nitrobacter. The uncoated urea was nitrified within 1-2 weeks but coated urea was hydrolysed in 4-6 weeks. There was no accumulation of nitrite in the soil.

379.
KHANNA, R.N. and SINGH, R.S. (1974) Rhizosphere populations of Fusarium species in amended soils.

Indian Phytopath. 27 : 331-339.

The growth and sporulation of Fusarium spp. in the rhizosphere of pigeon pea increased with plant age. In root-free soil, amendment with oil-cakes, sawdust and, to a lesser extent, urea stimulated Fusarium activity. In rhizosphere soil, this stimulation tended to be suppressed or reversed.

380.
KHANNA, USHA; SHETHNA, Y.I. and VYAS, S.R. (1972) Bacteriological activity in saline, saline-alkali and alkali soils. Curr. Sci., 40 : 269-270.

Total number of micro-organisms decreased logarithmically in the soil with the increase in pH. Maximum count of nitrogen fixing bacteria (8.5×10^2 per g of soil) was observed at pH 8.0 from Hissar soil and minimum (3.0×10^2 per g of soil) in Mahim soil of pH 10.3. The count of nitrogen-fixing bacteria was affected only slightly with pH.

381.
KHARE, H.P.; CHAUHAN, J.S. and TANDON, S.P. (1966) The effect of some rare elements on the growth of nitrite-formers.

Zentbl. Bakt. ParasitKde Abt. II, 120 : 117-122.

Small amounts of the chlorides of Zr and Ce and of the sulphates of In, Tl and Be accelerated the process of NO_2 formation by Nitrosomonas in culture media; an injurious effect, decreasing the formation of NO_2, occurred beyond the optimum concentration for each salt. Formation of NO_2 was decreased by OsO_4 at a concentration as low as 0.03177 mg/82 ml.

382.
KHARE, N.K. and RAI, M.M. (1968) Effect of phosphorus on symbiotic fixation of nitrogen by leguminous crops.

J. Indian Soc. Soil Sci. 16 : 111-114

The effect of phosphorus on symbiotic fixation of nitrogen by a few leguminous crops, namely, sunnhemp, soyabean, cowpea, dhaincha (Sesbania aculeata), moong (Phaseolus aureus) and urd (*Phaseolus mungo*) has been studied. The maximum amount of nitrogen was excreted by dhaincha followed by soyabean, cowpea ad moong. There was a negative balance of nitrogen in the cases of sunnhemp and urd. Leguminous crops when treated with phosphorus increased the nitrogen content of the soil significantly in the order, soyabean, dhaincha, urd, cowpea, moong and sunhemp. The total amount of nitrogen (plant and soil) fixed symbiotically in the phosphate treated pots over control was more than double in each case.

383.
KHATI, D.V.S.; BHARGAVA, S.N. and SHUKLA, D.N. (1983) Some studies on the saprophytic survival of *Sclerotium rolfsii* in soil.

Z. Pflant, Pflanzenschutz. 90 : 382-387

The survival of *S. rolfsii* was highest at soil moisture values of between 30 and 50% of water-holding capacity and at temperatures between 20 and 25°C. Addition of carbohydrates to the soil at a 1% rate caused a decline in the population on colonized wheat straw.

384.
KHURANA, A.L. and VYAS, S.R. (1977) Respiratory activity of bacteroids from bengalgram (*Cicer arietinum* L.) as a function of plant age.

J. Res. HAU. 7 : 139-141.

In trials with *C. arietinum* grown in sterilized soil given P. respiratory activity (RA) of bacteroids isolated from root nodules at 1-wk intervals beginning 1-wk after the appearance of the 1st nodule (20 days after sowing) was determined using different substrates. During the 8-wk of active life of root nodules, the RA of bacteroids changed according to a definite pattern. A depression occurred in RA at 31-41 days of plant etoglutarate, while it increased with the substrates succinate, fumarate and malate continued to increase up to a peak at 55 days after sowing; thereafter the RA declined steadily up to the age of 76 days.

385.
KOTHARI, S.K. and SARAF, C.S. (1984) Response of summer Mung (Vigna radiate L. Wilczek) to bacterial seed inoculation and phosphorus application. Golden Jubilee Seminar on Soil Resources and Productivity Management, Indian Society of Soil Science, New Delhi 3 : 58.

A pot culture experiment was conducted at the Indian Agricultural Research Institute, New Delhi, in summer 1983 with four types of bacterial inoculation (Rhizobium, Rhizobium along with Azotobacter chroococcum, Rhizobium with Azospirillum brasilense and no inoculation) and three levels of phosphorus (^{32}P labelled) application (0, 6.5 ppm, 13.0 ppm) in a factorial randomised design having three replicates. Inoculation with Rhizobium along with Azotobacter Chroococcum was found to be better than Rhizobium alone or Rhizobium with Azospirillum brasilense in respect of leaf, stem, root and nodule dry weight, phosphorus uptake and phosphorus utilisation efficiency. Leaf, stem root, nodule dry weight, per cent phosphorus derived from fertilizer (% pdff) and fertilizer P uptake increased, while phosphorus utilisation efficiency decreased with each increment in the level of applied phosphorus. Under P stress condition (control) nodule growth was inhibited relatively more than that of either root or shoot.

386.
KOTHARI, S.K. and SARAF, C.S. (1985) Effect of bacterial seed inoculation and phosphorus application on growth, yield, uptake and quality of summer mung (Vigna radiata L. Wilczek).

Proc. Natn. Sem. Plant Physiol. (Abst.) p. 39

A field experiment was conducted at IARI New Delhi to study the effect of heat tolerant strains of bio-fertilizers and phosphorus on growth and yield of summer mung. The bacterial seed inoculation treatments comprised no bacterial inoculation, inoculation with efficient and heat tolerant rhizobium strain; rhizobium strain + Azotobacter brasilense and a cultivated fallow. Seed inoculation with efficient strain of rhizobium along with Azotobacter Chroococcum; rhizobium strain + Azotobacter chroococcum gave consistently better performance than inoculation with rhizobium alone or rhizobium along with A. brasilense. Bacterial inoculation resulted in 10 to 17 per cent increase in grain yield.

387.
KOTWAL, Y.N. (1927) Studied in intensive bacterial oxidation. III - The nitrification of ammonia in strong solutions. J. Indian Inst. Sci. 10A : 102-106.

No appreciable loss of ammonia from the solution could be found when the air drawn through the cylinders was passed through decinormal sulphuric acid and the acid tested for the presence of ammonia at the end of a month.

388.
KRISHNA, K.R. and BAGYARAJ, D.J. (1984) Growth and nutrient uptake of peanut inoculated with the mycorrhizal fungus Glomus fasciculatum compared with non-inoculated ones. Pl. Soil 77 : 405-408.

In a greenhouse study, inoculation with the mycorrhizal funqus Glomus fasciculatum enhanced peanut growth and increased its dry matter more than two fold compared with the non-inoculated control, in both sterile and non-sterile soil. It also significantly increased uptake of phosphorus and micronutrients such as zinc, copper, manganese and iron.

389.
KRISHNAMURTI, B.G. (1962) Nitrogen fixation with molasses. Indian Sug. 11 : 765-767.

Molasses applied to moist soil supplied carbohydrate for utilization by N-fixing soil bacteria. Addition of $CaCO_3$ increased fixation; CaO was also effective at low concentration.

390.
KUKEWAR, P.M. and BHIDE, V.P. (1969) Effect of seed treatment with fungicides and antibiotic aureofungin on the efficacy of nodulation by Rhizobium strain of groundnut.

Hindustan Antibiot. Bull. 11 : 172-176

All the *Rhizobium* spp. tested could grow well in the presence of Agrosan, Brassicol and Aureofungin. However, growth of *R. phaseoli, R. japonicum, R. trifolii* and *R. meliloti* was inhibited by Thiram. Captan inhibited growth of *R. leguminosarum* but not of other *Rhizobium* spp. Showing that different species react in different manner towards these fungicides. Brassicol and aureofungin appear to be choice material for treating the seeds.

391.
KULKARNI, J.H.; SARDESHPANDE, J.S. and BAGYARAJ, D.J. (1974) Effect of four soil applied insecticides on symbiosis of *Rhizobium* sp. with *Arachis hypogaea* Linn.

Pl. Soil. 40 : 169-172.

Application of carbofuran, thimet, dasanite and heptachlor at the recommended rates had no harmful effect on the nodulation and yield of groundnut inoculated with *Rhizobium.*

392.
KULKARNI, S.; SIDDARAMAIAH, A.L. and BASAVARAJAIAH, A.B. (1980) Antagonistic action of *Streptomyces* sp. on *Rhizoctonia bataticola,* a common soil borne pathogen of forest nurseries. Indian Forester 106 : 126-127.

In *in vitro* tests a *Streptomyces* sp. exhibited strong antagonism towards *R. bataticola* (*Macrophomina phaseolina*).

393.
KUMAR, H.D. and LADHA, J.K. (1976) Comparative studies of growth and nitrogen fixation in certain subclones of the cyanophyte *Glococapsa.*

Biochem. Physiol. Pflanzen. 169 : 63-69.

No marked differences in growth and nitrogen fixation were observed between the seven subclones studied although low level significant differences were observed between mean cell lengths of a fast-growing and slow-growing subclone. The observed variability may be evidence of the existence of genetically distinct clonal or subclonal populations of the alga in natural habitats.

394.
KUMAR, H.D. and SINGH, H.N. (1977) Studies on the algae of certain habitats endemic to Rajasthan. In The Natural Resources of Rajasthan (edited by M.L. Roonwal), University of Jodhpur, Jodhpur, 1 : 65-78

A general study of the algal flora of certain habitats endemic to Rajasthan has been made in relation to some physicochemical factors with a view to studying the tolerance range ecology of Rajasthan algae and to assess their role in the biological productivity of their respective habitat.

395.
KUMAR, H.D. and SINGH, H.N. (1978) Physiology and genetics of nitrogen-fixation in blue-green algae. pp. 287. In S.P. Sen, Y.P. Abrol and S.K. Sinha (eds.), Nitrogen Assimlation and Crop Productivity. Associated Publishing Company, New Delhi.

The regulation of nitrogen fixation and photosynthetic CO_2 fixation in blue-green algae shows a basic difference in that whereas the nitrogen-fixing apparatus is repressible by ammonia or other form of fixed nitrogen, the photosynthetic CO_2 fixing apparatus lacks such a repression-depression control mechanism. Heterocyst formation and nitrogenase synthesis show significant level of coordinate repression and derepression. Our studies on the behaviour of some classes of non-nitrogen-fixing (Nif-) mutants suggest that heterocyst and nitrogenase regulation might be controlled by a common positive regulatory gene. This paper presents genetic evidences to show that heterocyst differentiation is under positive control and heterocyst pattern formation is under negative control in some species of *Nostoc* and *Anabaena*.

396.
KUMAR, N. and BANERJI, D. (1978) Interference in the nitrogen

metabolism of root nodules of <u>Cicer arietinum</u> (bengal gram) by lead. pp. 293

<u>In</u> S.P. Sen, Y.P. Abrol and S.K. Sinha (eds.), Nitrogen Assimilation and Crop Productivity. Associate Publishing Company, New Delhi.

During studies on the effect of polluted irrigation water on crop plants, lead was detected as the major heavy metal pollutant in the irrigation water, irrigated soil and parts of plants grown in such polluted soil. Further studies were, therefore, undertaken to investigate the biochemical effects of lead on crop plants.

Lead was added in the soil and as Pb. $(NO_3)_2$ to study its effects on nitrogen fixation in Bengal Gram. Plants grown in lead containing soil have lower levels of protein nitrogen and soluble nitrogen in root nodules. The amount of protein in the nodules of control sets is 33.7 mg/gm fr.wt. and 30.8 mg/gm fr.wt. in experimental sets. Soluble nitrogen level also showed the same pattern i.e. 3.6 mg/gm fr.wt. and 3.0 mg/gm. fr.wt. for control and experiemental sets, respectively. It appears that lead has inhibitory effect on the process of nitrogen fixation by root nodules.

397.
KUMAR, S.; SHARMA, H.R. and GARG, O.P. (1981) Effect of saline conditions on nodulation and nitrogen fixation by root nodules of <u>Pisum sativum</u> L (pea). <u>Indian J. Microbiol.</u>, 21: 211.

Pea plants were grown in sand culture maintained at four conductivity levels viz. 2.1, 6.0, 9.0 and 12.0 mmhos/cm (prepared by the addition of NaCl, $CaCl_2$ and $NaHCO_3$ to a standard N-free nutrient solution). Salinity depressed nodulation and nitrogen fixation. The number, size and weight of the nodules were all progressively reduced by increasing levels of salinity. However, the growth of rhizobial culture was not inhibited by these levels of salinity. Salinity delayed the development of leghaemoglobin in the nodules, reduced its concentrations, though slightly, and, towards later growth stages, enchanced the greening of this pigment.

398.
KUMARA RAO, J.V.D.K.; SEN, A. and SHENDE, S.T. (1974) Inhibition of groundnut *Rhizobium* in Indian soils.

Proc. Indian natn. Sci. Acad. 40B : 535-539.

Introduction of strains of groundnut *Rhizobium* in soil by way of inoculation is associated with competition not only from indigenous *Rhizobium* but also antagonism from some bacteria, fungi and actinomycetes. Soluble constituents of soils also invariably exert a depressing effect on the growth of the *Rhizobium*.

399.
KUNDU, B.S. and GAUR, A.C. (1981) Inoculant carriers for phosphate-solubilizing micro-organisms.

Indian J. agric. Sci. 51 : 252-255

Peat, sandy loam farm soil, farmyard manure and cowdung-cake powder were examined for their suitability as carriers for phosphate-solubilizing micro-organisms. Peat was the best carrier for survival and multiplication of *Bacillus polymyxa, Pseudomonas striata* and *Aspergillus awamorii* up to 60 days after inoculation for single and mixed inoculants. The sporulating bacilli and *Aspergillus* also survived well in the soil.

400.
KUNDU, B.S.; DADARWAL, K.R. and TAURO, P. (1982) Effectivity of cross-infective *Rhizobia* in heterologous legume hosts *in vivo*.

Zentb. Mikrobiol. 137 : 9-13

Nodulation and nitrogen-fixing ability of cross-infective strains of *Rhizobium japonicum* and *Rhizobium* sp. 'cow pea group' were tested in different hosts of two groups. In distantly related hosts the cross nodulating bacteria were often non-effective in nitrogen fixation. Therefore, *Rhizobium* strain, genetically manipulated for wide host infectivity, cannot be of much practical advantage in use as legume inoculant for hosts of more than one cross inoculation group, unless the nitroge-fixing ability of the bacterium is also manipulated.

401.
KUSHWAHA, R.K.S. and AGRAWAL, S.C. (1976) Some keratinophilic fungi and related dermatophytes from soil.

Proc. Indian natn. Sci. Acad. 42 B : 102-110.

Soil samples collected from different localities frequented particularly by various animals and birds were baited for the isolation of fungi capable of colonizing and attacking keratinous substratas. In all, 94 fungi were isolated. The distribution of keratinophilic fungi in various types of soils and on different kinds of keratinic substrates is discussed. Some fungal species new to Indian fungal flora are mentioned.

402.
LADHA, J.K. and KUMAR, H.D. (1977) Response of a wild-type and a non-nitrogen fixing mutant of Nostoc linckia to different carbon sources.

New Phytol. 79 : 299-308.

The wild-type parentalstrain and a het+nif- mutant of Nostoc linckia grew heterotrophically in light, with and without DCMU, and in darkness when supplemented with a source of carbon. Glucose and sucrose stimulated heterocyst production. Glucose enhanced the uptake of ammonium-N when parent had mutant were grown in an ammonium-supplemented medium. Effects of cAMP (sodium salt and dibutyryl), with and without glucose, on growth and heterocyst differentiation of parent and mutant were studied. Dibutyryl cAMP was generally toxic, but the sodium salt enhanced growth with glucose. The mutant grew well in a nitrogen-free medium supplemented with cAMP.

403.
LADHA, J.K. and KUMAR, H.D. (1978) Characterization of some

mutants of *Nostoc linckla* with respect to nitrogen fixation and heterocyst formation. pp. 290-291. *In* S.P. Sen, Y.P. Abrol and S.K. Sinha (eds.), Nitrogen Assimilation and Crop Productivity. Associated Publishing Company, New Delhi.

Several classes of mutants of a wild-type *Nostoc linckia* have been isolated, viz.: (1) not fixing nitrogen but producing heterocysts ($het^{+}nif^{-}$), (2) neither fixing nitrogen nor producing heterrocysts ($het^{-}nif^{-}$) and (3) fixing nitrogen and producing heterocysts in an altered spacing pattern.

404.
LAHIRI, K.K. (1974) Differential competition among the strains of *Rhizobium japonicum* for nodule sites on soyabean. *Proc. Indian natn. Sci. Acad.* 40B : 644-649.

In experiments conducted with eight serologically distinct strains of *Rhizobium* and twelve different varieties of soyabean, it was observed that the differential competition between the strains of *Rhizobium* in nodulating soyabean varieties is the resultant expression of two interacting forces operating simultanaously, viz. (i) preferentiality exhibited by a particular soyabean variety to an inoculating *rhizobial* strain, and (ii) the extent to which such imposed preferentiality may be exploited by the particular bacterial strain. The magnitude of the second component depends more on the nature of the inoculating strains and could be manipulated to some extent artificially. In the light of the results the phenomenon of competition among the strains of *Rhizobium* for nodule sites has been critically examined.

405.
LAKSHMANARAO, K. and KALYAN SINGH (1983) Effect of phosphorus and bio-fertilisers on leghaemoglobin and nitrogen fixation of chickpea.

Madras agric. J. 70 : 572-577.

Results show that the fresh weight of nodules increased significantly with increasing levels of phosphorus as well as the Rhizobium inoculation. Rhizobium inoculation had marked influence on leghaemoglobin content of nodule tissue. The total leghaemoglobin content in the nodules is directly related to the total nitrogen fixed by the chickpea.

406.
LAKSHMI KUMARI, M. (1964) Rhizosphere microflora of Dolichos infected with DEMV. Proc. Indian Acad. Sci. 60B : 116-127

The effect of mosaic virus infection on rhizosphere microorganisms of Dolichos lablab are discussed. There is a decrease in bacterial and actinomycete numbers and increase in fungi.

407.
LAKSHMI KUMARI, M.; BISWAS, A.; VIJAYALAKSHMI, K.; NARAYANA, H.S. and SUBBA RAO, N.S. (1974) Effect of certain water-soluble herbicides on legume - Rhizobium symbiosis. Proc. Indian natn. Sci. Acad. 40B : 528-534.

Among the water-soluble herbicides tried dalapon was inhibitory to the growth of rhizobia but the herbicide had no noticeable effect on nodulation of lucerne under bacteriologically controlled conditions. Both dalapon and amitrole induced the formation of pseudonodules on the roots of legumes even in the absence of Rhizobium.

408.
LAKSHMI KUMARI, M.; SINGH, C.S. and SUBBA RAO, N.S. (1974) Root hair infection and nodulation in lucerne (Medicago sativa L.) as influenced by salinity and alkanity. Pl. Soil, 40: 261-268.

Experiments with lucerne under bacteriologically controlled conditions on the effects of chloride, carbonate and bicarbonate of sodium, alkaline pH levels and saline/alkaline soil extracts demonstrated that increasing levels of salinity/alkalinity of the root medium resulted in a root system devoid of root-hairs, mucilaginous layer and infection thread formation inspite of the optimum growth of Rhizobium under such conditions. The results point out the existence of 'alkali sensitive steps' in the early phases of nodulation which lead to an elimination of legume rhizosphere as postulated by earlier workers.

409.
LAKSHMI, V.; SATYANARAYANA RAO, A.; VIJAYALAKSHMI, K.; LAKSHMI KUMARI, M.; TILAK, K.V. and SUBBA RAO, N.S. (1977) Establishment and survival of *Spirillum lipoferum*.

Proc. Indian Acad. Sci. 86B : 397-405.

S. lipoferum able to fix atmospheric N in culture medium was isolated from the roots of several plant species. In seedlings inoculated with *S. lipoferum*, the bacteria occurred in root hairs, root epidermal cells and very occasionally root xylem. In the laboratory, the addition of farmyard manure to soil inoculated with *S. lipoferum* and then incubated at 28°C increased the survival time of the bacteria from 1 week to a maximum 31 weeks. *S. lipoferum* was isolated from air-dry soil that had been stored for 15 days.

410.
LAKSHMINARASIMHAM, A.V. and RAO, P.R. (1969) Three new records of fungi from soils.

Curr. Sci. 38 : 73-74.

Acrospeira fluctuata, *Helminthosporium hawaiiense* and *Pithomyces sacchari* were isolated respectively from soil supporting *Anacardium occidentale*, coastal dune soil, and soil supporting *Eucalyptus* sp.

411.
LAL, A. and ACHARI, T.K.T. (1953) Microbiological studies VII - Studies on *Azotobacter* and Actinomycetes in relation to nitrogen fixation and cellulose utilization. *Proc. nat. Acad. Sci. India* 23 : 137-149.

The azotobacter population in Indian soils is lowest in winter and increases during the rains. There is a definite ratio of azotobacter to other organisms in many soils and a positive correlation with soil pH. Actinomycetes in association with azotobacter fixed more N than either organism alone, the maximum being reached after 21 days of incubation, and fixation being promoted by the presence of humus.

412.
LAL, A. and DUBE, H.D. (1953) Microbiological studies. IV - Decomposition of horse manure. Proc. nat. Acad. Sci. India 23 : 93-101.

In pot experiments involving treatment of horse manure with various N fertilizers, decomposition of cellulose was higher under aerobic than anaerobic conditions. Lignin was more resistant than other constituents. Anaerobic conditions resulted in a decrease of N and increase of C : N ratios. Incorporation of N fertilizers, especially NO_3 increased microbial activity, hastened decomposition and resulted in almost complete cellulose decomposition under aerobic conditions.

413.
LAL, A. and MUKERJI, S.K. (1953) Microbiological studies. II - Carbon dioxide evolution and microbial activities in the soil under different manurial treatments. Proc. nat. Acad. Sci. India 23 : 66-74.

In brown light loam soils, treatments with 0.15% dextrose was associated with the highest CO_2 production and microbial numbers. Urea increased and ammonium sulphate decreased CO_2 production. Microbial activity and CO_2 production were depressed by mixtures of soil, NO_3 fertilisers and 10% manure and by NO_3 fertilizers at rates $>$ 0.2%. FYM ripened for six months yielded highest CO_2 and bacterial count.

414.
LAL, R. and MISHRA, M.M. (1978) Cellulolytic activity of some soil fungi.

Folia Microbiol. 23 : 68-71.

Four species of Aspergillus, two of Fusarium, three of Paecilomyces and one each of Masoniella, Monosporium, Rhiozoctonia and Stachybotrys, isolated from soil and compost pits, were tested for cellulolytic activity. The cellulase produced in the presence of cellulose powder was active against carboxymethyl cellulose (CMC), and that produced in the presence of CMC was active against cellulose powder.

415.
LAL, V.B.; SINGH, J. and PRASAD, B. (1981) Hemifridericia varanensis, a new species of Enchytraeidae (Oligochaeta) from tropical soils of estern Uttar Pradesh, India.

Rev. Ecol. Biol. Sol. 18 : 263-267

Hemifridericia varanensis sp. n. (Oligochaeta, Enchytraeidae) from eastern Uttar Pradesh, India, is described. The genus Hemifridericia is reported from India for the first time.

416.
LAURA, R.D. (1977) Salinity and nitrogen mineralization in soil.

Soil Biol. Biochem. 9 : 333-336.

A laboratory study determined the effects of salinity on ammonification, nitrification and mineral N accumulation in incubated soils. Ammonium-N, nitrite-N and nitrate-N were measured periodically for 102 days in unamended soil of varying salinity and in soil amended with farm compost, mustard oil cake or urea. Increased salinity progressively retarded ammonification but did not suppress it completely. Nitrification was retarded, suppressed or inhibited completely by salinity, the effect depending on both the amount of salt and the type of amendment. The amount of mineral N that accumulated generally decreased with increased salinity.

417.
LEELAVATHY, K.M. (1969) Effect of rhizosphere fungi on seed germination.

Pl. Soil 30 : 473-476.

Germination of Dichanthium annulatum, Bothriochra fertusa and Setaria glauca seeds was not significantly affected by inoculation with different fungal filtrates, except in the case of D. annulatum inoculated with Trichoderma viride filtrate.

418.
LEHRI, L.K. and MEHROTRA, C.L. (1968) Use of bacterial fertilizers in crop production in U.P.

Curr. Sci. 37 : 494-495.

A few data on the effects of phosphobacterin or azotobacterin on yields, mainly of cereals, are given.

419.
LEHRI, L.K. and MEHROTRA, C.L. (1972) Effect of Azotobacter inoculation on the yield of vegetable crops.

Indian J. agric. Res. 6 : 201-204.

Yield increases of up to 50% in cabbage and 62% in brinjal were obtained after inoculation with Azotobacter strains. The increases varied both with the fertility status of the soil and with the different strains tested.

420.
LEHRI, L.K. and PRASAD, S.N. (1976) In vitro inhibitory effect of some fungicides on Rhizobium spp. and Azotobacter chroococcum.

J. Indian Soc. Soil Sci. 24 : 214-216

A study to find out the effect of some fungicides (Dithane M 45, Dithane Z 78, Flit 406, Ceresan wet and Aratan at 1500, 2000 and 3000 ppm conc.) on important Rhizobium spp. and on Azotobacter chroococcum. Azotobacter chroococcum showed considerable variation in its tolerance to these fungicides. This microbe turned out to be resistant to Dithane M-45 at all the tested concentrations and to Dithane Z-78 at 1500 ppm level. Flit 406, Ceresan and Aratan were inhibitory at this organism at all the tested concentrations especially at higher concentrations. It is further evident that all the legume seed dressing fungicides tested are inhibitory to Rhizobium spp. in vitro. This necessitates adequate care to protect the Rhizobium spp. from direct contact with dressant chemicals. It is felt that a prior coating of the chemically treated seeds with some pelleting materials like basic slag or ground lime may be useful to minimize their inhibitory effect on inoculated Rhizobium spp.

421.

LEHRI, L.K., GANGWAR, B.R. and MEHROTRA, C.L. (1974) Bacterization experiments with rhizobium.

J. Indian Soc. Soil Sci. 22 : 66-69

Out of 24 rhizobium strains screened for gram (Cicer arietinum), 9 are found to be markedly effective giving significant increase in yields of legume crops over uninoculated ones. Best response in urd (Phaseolus mungo) and lobia (Vigna catjang) are obtained through the use of efficient strains (A.C.) isolated and selected after the plant testing on their homologous hosts in this laboratory. Rhizobium strains (A.C.) gave about 50 per cent additional yield of urd and lobia in presence of 80 kg P_2O_5/ha. The responses were best exhibited at lower level of fertility in T4 and T49 varieties.

422.

LODHA, B.K. and LAL, P.B. (1969) Effect of pesticide treatments of seeds on the rhizosphere Microflora of some field crops.

J. Indian Soc. Soil Sci. 17 : 75-78

Barley, wheat and pea plants were grown after treating the seeds with Lindane, Endrin and Agrosan GN. The effects of the pesticides on the rhizosphere microbiological make-up have been variable. At low levels of application in certain instances, stimulation led to increased microbiological counts, but at higher levels of application a depression in counts was recorded.

423.
MADHAVA, RAO, A.R.; KATTI, V.M.; PANCHAKSHARAIAH, S.; RAO, K.B. and KUMAR, E.R.S. (1979) Effect of saline water on chemical properties and microbiological population in black soil. Mysore J. agric. Sci., 13 : 262-264.

The effect of application of saline water of varying concentrations to black soil over four years on some chemical properties and microbial population, indicated that pH and ESP increase with increasing salinity. The microbial counts tend to increase in the lower salinity ranges; but they decrease at higher salinity levels. Similar parallel variation in the organic carbon content of the treated soil is observed. Actinomycetes appear to be more tolerant to salinity and sodicity than bacteria and fungi.

424.
MAGU, S.P.; GAUR, A.C. and SADASIVAM, K.V. (1974) Effect of lindane on nodulation, grain yield and nitrogen uptake of gram (Cicer arietinum). Proc. Indian natn. Sci. Acad. 40B : 526-527.

Lindane at 1 ppm was not toxic to nodulation or plant growth, but 5 ppm and above inhibited nodulation, grain weight and nitrogen uptake. At 10 ppm, nodulation was completely suppressed. At 100 ppm, germination was considerably reduced.

425.
MAHAPATRA, I.C.; MUDHOLKAR, N.J. and PATNAIK, H. (1971) Effect of N, P and K fertilizers on the prevalence of algae under field conditions.

Indian J. Agron. 16 : 19-22.

P at rates of 40, 80 and 120 kg/ha P_2O_5 increased algal growth in rice fields. N had a lesser effect but rates of up to 80 kg/ha N also favoured algal growth. The effect of N + P was additive, but K alone or in combination with N and P decreased growth. Chroococcus and Volvox were the dominant genera, and these may compete with the rice crop for N and P. Additional K would prevent algal mat formation and increase the availability of N and P to high yielding rice varieties.

426
MAHAPATRA, R.N. and RAO, V.R. (1981) Influence of hexachlorocyclohexane on the nitrogenase activity of rice rhizosphere soil.

Pl. Soil 59: 473-477

Application of HCH to a submerged paddy field soil at rates of up to twice recommended field rates (1-5 kg a.i./ha) stimulated nitrogenase activity and increased populations of Azospirillum sp., Azotobacter sp. and, to a lesser extent, anaerobic N_2 fixers. HCH retarded the drop in redox potential of the soil for up to 80 days after transplanting.

427.
MAHAPATRI, B.; PATNAIK, B. and MISHRA, D. (1977) The exocellular urease in rice roots.

Curr. Sci. 46: 680-681.

Urea hydrolysis by exocellular urease bound to the surface of rice roots, and by urease released from the roots into the nutrient solution suggests that these exocellular ureases may hydrolyse organic substances in soil independently of soil microbial activity.

428.
MAHATIM SING, JAGDISH SINGH and KALYAN SINGH (1983) Effect of phosphorus and biofertilizers on chlorophyll content of leaves and leghaemoglobin contents of fresh nodules in Kharif grain legumes.

Indian J. Agron. 28 : 229-234.

Chlorophyll content in leaves as well as nodule-dry weight were significantly influenced by phosphorus as well as bio-fertilizers in grain legumes *viz* black gram, cowpea and green gram. Leghaemoglobin content was also markedly influenced due to the use of bio-fertilizers during both the years of experimentation (1979 & 1980). Chlorophyll and leghaemoglobin content were positively associated with the grain yield of legumes.

429.
MAHESHWARI, R. (1968) Occurrence and isolation of thermophilic fungi.

Curr. Sci. 37 : 277-279.

Malbranchea pulchella, Talaromyces thermophilus, three varieties of *Chaetomium thermophile* and two species of *Humicola* were isolated from dung, compost and sewage manure. A third species of *Humicola* was found in the samples but not isolated in culture.

430.
MALL, S. (1979) Rhizosphere and rhizoplane microflora of three potato varieties.

Indian Phytopath. 32 : 51-54

Quantitative and qualitative studies on the rhizosphere and rhizoplane microflora of three potato varieties showed that the microflora increased with age of the plant and reached a maximum in 6-8 week-old-plants. For bacteria and actinomycetes the greatest rhizosphere effect was for variety Local and for fungi it was for variety Up-to-date. Both rhizosphere and rhizoplane microflora of the moderately resistant variety Up-to-date were characterized by the continuous presence of *Trichoderma viride* at high frequency as compared to the other two varieties. The rhizoplane microflora was dominated by sterile forms and appeared to be a better index of the effects of roots on microbes than the rhizosphere.

431.
MALLIKARAJUNAIAH, R.R. and BHIDE, V.P. (1981) Effect of

Azotobacter on germination of spores of different fungi.

Zentbl. Bakt. ParasitKde. Abt. II, 136 : 563-565

The fungistatic effect of Azotobacter cultures on spore germination of Alternaria (5 cultures), Curvularia (5 cultures), Corynespora (1 culture), Fusarium (3 cultures) and Helminthosporium (4 cultures) was studied. Germination percentages were recorded after a period of 6 hours at 30°C.

432.
MALLIKARAJUNAIAH, R.R. and BHIDE, V.P. (1982) Fungistatic action of Azotobacter chroococcum and A. vinelandii on different fungi.

Zentb. Mikrobiol. 137 : 55-62

The sixty fungi tested, belonging to pathogenic and non-pathogenic genera, interacted with Azotobacter in both directions, viz, stimulation as well as inhibition of growth. Most of the Azotobacter cultures of both the species were active against Alternaria, Curvularia, Fusarium, Rhizoctonia, Helminthosporium, and Verticillium. At the same time 17 out of 20 cultures of Azotobacter were inhibited by Aspergillus sp. (PS-1) and all the 20 cultures by Aspergillus sp. (KR-31). Penicillium commune (D-16) inhibited the growth of all Azotobacter cultures; three cultures of Trichoderma (S-1, No. 2, and M-2) had inhibitory effect on all the 20 cultures of Azotobacter.

433.
MANDAL, L.N. (1984) Soil research in retrospect and prospect. III-Soil biology and biochemistry. In Soil Science in India.

Bull. Indian Soc. Soil Sci. 14 : 124-135

The article provides a brief account of the state of research in the field of soil biology and biochemistry and give a projection of future line of research. A good deal of work has been done in this country on microbial transformation of soil as well as applied organic nitrogen, particularly on the ammonification and nitrification processes and the various environmental factors influencing the rate of the above processes. The article throughs light on biological nitrogen fixation, microbiological decomposition of organic matter, etc.

434.
MANJUNATH, D.J.; BAGYARAJ, D.J. and GOPALA GOWDA, H.S. (1984) Dual inoculation with VA mycorrhiza and rhizobium is beneficial to Leucaena. Pl. Soil 78 : 445-448.

Response of Leucaena leucocephala to inoculation with Glomus fasciculatur and/or Rhizobium was studied in a phosphorus deficient unsterile soil. G. fasciculatum only inoculation improved nodulation by native rhizobia and Rhizobium only treatment improved colonization of roots by native mycorrhizal fungi. Dual inoculation with both the organisms improved nodulation, mycorrhizal colonization, dry weight, nitrogen and phosphorus content of the plant compared to single inoculation with either organism.

435.
MANOHARACHARY, C. (1977) Microbial ecology of scrub jungle and dry waste land soils from Hyderabad district, Andhara Pradesh (India).

Proc. Indian natn. Sci. Acad. 43 : 6-18.

In both soils the fungal population decreased in summer whereas the bacterial population was greatest in summer. In the waste land soil but not in the scrub jungle soil, the number of fungi was positively correlated with the pH and moisture content of the soil. In the scrub jungle soil, the number of fungi was positively correlated with the organic matter content. The fungal population consisted of 112 species, of which 67 occurred in both soils.

436.
MARATHE, K.V. (1965) A study of the subterranean algal flora of some mangrove swamps.

J. Indian Soc. Soil Sci. 13 : 81-84

The subterranean algal flora of five salt marshes around Bombay were small in number. The higher percentage of chloride and of clay and silt are the factors responsible for this small number. Out of the twelve algal species obtained, ten belong to Cyanophyta and two to Chlorophyta. This suggests the adaptability of some members of Cyanophyta to saline soils.

437.
MARATHE, K.V. and CHAUDHARI, P.R. (1975) An example of algae as pioneers in the lithosphere and their role in rock corrosion.

J. Ecol. 63 : 65-69.

The role of an Aulosira aenigmatica community in rock corrosion and soil formation is described. The results indicate that this community had a considerable corroding action and led to corrosion of the rocky substratum equivalent to fifteen times the weight of the alga. The algal community also had a high moisture-absorbing power, absorbing water almost to its weight.

438.
MATHAN, K.K., RAMASWAMI, P.P. and NAIR, K.S. (1976) Seasonal variations in the distribution of Clostridium sp. in red and black soil profiles.

J. Indian Soc. Soil Sci. 24 : 334-335

In black soils the Clostridium population increased from winter to spring by 57.2% while from spring to summer the increase was 34.6% and from summer to autumn it was only 0.2%. The population decreased by 52.8% from autumn to winter. In red soil, the increase in population from winter to spring was 76.7%. Then the population decreased by 15.2% towards summer and from summer it decreased by 54.9% towards autumn. From autumn to winter the increase in population was 47.9%. From winter to spring there was the maximum increase and that was more pronounced in red soil. Further the total population of Clostridium was more in black soil than in red soils. Reduction in population from surface horizon to the bottom most horizon was more pronounced in black soil than in red soil.

439.
MATHAN, K.K.; RAMASWAMI, P.P.; SUYAMBULINGAM, C. and JEBARANI, W. (1977) Indentification of factors influencing microbial population in red and black soils by path analysis.

Madras agric. J. 64 : 549-553.

Moisture, pH, depth and carbon dioxide evolution were among the major factors influencing the microbial population in red and black soil. Path analysis revealed that the net contribution of pH and depth was negative and moisture and carbon dioxide evolution positive. The effect of pH was mostly indirect, via depth, in black soil, and direct in red soil. The contribution of moisture was direct in both soils. The direct effect of CO_2 was low and negative in both soils. Depth had a direct negative influence on population in both soils.

440.
MEHROTRA, C.L. and LEHRI, L.K. (1970) Effect of Rhizobium Inoculation in Legume Crops.

J. Indian Soc. Soil Sci. 18 : 203-207

The studies on isolation and selection of root nodule bacteria of major legume crops of Uttar Pradesh, viz., berseem (Trifolium alexandrinum) and pea (Pisum sativam) in winter and dhaincha (Sesbania aculeata), sanai (Crotolaria juncea), urd (Phaseolus mungo), moong (Phaseolus aureus), lobia (Vigna catjang) and Soyabean (Glycine max) rainy season crops have given extremely encouraging trends and a large number of effective strains of these crops have been isolated. The effectiveness of strains were initially tested in the laboratory by recent Gibson's technique and subsequently in the field. Inoculation of rainy season legumes with efficient strains has shown very promising effect even in tropical soils.

441.
MEHROTRA, C.L. and LEHRI, L.K. (1971) Effect of Azotobacter inoculation on crop yields.

J. Indian Soc. Soil Sci. 19 : 243-248.

In field experiments wheat and rice fertilized with up to 150+75 +75 kg/ha N+P +K and cabbage, tomato and brinjal fertilized with NPK and FYM were inoculated with Azotobacter strains isolated from various soils in Uttar Pradesh. Inoculation did not significantly increase wheat, rice, brinjal or tomato yields but cabbage yields increased significantly after inoculation with some of the Azotobacter strains.

442.
MEHTA, A.R.; LODHA, B.K. and PUNTAMKAR, S.S. (1980) Note on the effect of salts on the rate of nitrification of nitrogenous fertilisers. Curr. Agric., 4 : 107-109.

The rate of nitrification was more in case of ammonium sulphate as compared to urea, and increasing concentrations of salt reduced the rate of nitrification.

443.
MEHTA, B.V., PATEL, G.J. and DANGARWALA, R.T. (1963) Effect of fumigation and other measures for nematode control on the production of nitrate in Goradu soil.

J. Indian Soc. Soil Sci. 11 : 361-371.

Nitrate contents of goradu soil, treated with four different measures viz., (i) irrigation and complete removal of stalks with roots, (ii) spot fumigation in summer (SFS), (iii) complete fumigation in summer (CFS) and (iv) spot fumigation in monsoon (SFM) with D.D. for the control of nematodes in presence of five different fertilizers, were determined under field conditions at different stages of growth of tobacco plant. It was found that fumigation in general, had a depressing effect on the formation of nitrate at the transplanting stage, this effect being maximum in the case of SFM treatment. In decreasing order of nitrate formation at the transplanting stage, the different control measures can be arranged as : control > irrigation > SFS > CFS > SFM. But at the topping stage the nitrate content under different fumigation treatments was not less than under no fumigation. Though fumigation brought about a temporary retardation of nitrification, the nitrogen nutrition of the plants did not suffer.

444.
MEHTA, P.M. and ABHISWAR SEN (1957) Studies on biological fixation of nitrogen in some Indian soils.

<u>J. Indian Soc. Soil Sci.</u> 5 : 189-197

Nitrogen fixations in twenty surface soils from different parts of India with widely varying pH, carbon, nitrogen and total phosphoric acid contents were determined in mannite solution with minerals under different conditions.

It was observd that there might be a wide deficiency of phosphates in Indian soils as there were increases in nitrogen fixation by the mixed microbiological population on addition of soluble phosphates like dipotassium hydrogen phosphate in most of them.

Under similar supply of soluble phosphate, nitrogen fixation in the soils depended not only on the nature of the soil and microbiological population, but also on the number of the microorganisms involved.

Activity of the nitrogen fixing microbiological population of a soil under optimum conditions of phosphate supply did not appear to be influenced by pH, organic carbon, organic nitrogen or the total phosphate content of the soil. Lime seemed to play a rather unimportant role in fixation of nitrogen in soils.

445.
MIAH, M.A.J.; VARSHNEY, J.L. and SARBHOY, A.K. (1980). Soil fungi of South India. <u>Proc. Indian natn. Sci. Acad.</u> 46B: 593-602.

Two hundred and fifty one different fungi were identified. Maximum numbers of fungi were isolated by the soil plate method. Basidiomycetes and Mycelia sterilia were absent. No definite correlation was established for the distribution of fungi in relation to C/N ratio.

446.
MIDHA, S.K. and NANDWANA, R.P. (1974) Effect of certain nematicides on the population of soil micro-organisms. I. Quantitative and qualitative changes in the fungal flora.

Indian Phytopath. 27 : 312-315.

A reduction in the fungal population of the soil was observed ten days after application of DD, ethylene dibromide, DBCP, Dasanit and Nemafos to field plots; at sowing time, 25 days after application, the fungal population was increased to levels two to five times higher than that at pre-application levels. By harvest, 145 days after application, the fungal population showed increases of 9 to 50% over untreated plots. The differential susceptibilities of Aspergillus, Cladosporium, Fusarium, Curvularia and Myrothecium to the chemicals is reported.

447.
MISHRA, A.S. and VYAS, K.M. (1972) Production of α- proline by soil micro-organisms.

Hindustan Antibiot. Bull. 15 : 30-33.

Out of 32 fungi only 17 species were found to produce α proline in their culture filterates. In investigations of various media and conditions, it was found that maximum amount of α-proline was produced in a medium containing 3% glucose.

448.
MISHRA, D. and MOHANTY, B. (1967) External effect of indolyl--3-acetic acid, benzimidazole and Bnine on nodulation of cowpea.

Nature, 214 : 320-321.

The theories of (1) internal effect of growth retardants (on development of nodule) and (2) external effect (on infection) appear complementary and compatible. The present investigation suggests that there is such a local external effect of the retardants in inhibiting nodulation. Whether there is an inhibition in the infection process or a reduction in the number of sites of infection awaits further experimentatirn.

449.
MISHRA, K.B. (1979) Effect of soil amendment with root extract of Hibiscus cannabinus L. on its rhizosphere mycoflora.

Pl. Soil. 51 : 421-424.

Hibiscus cannabinus was grown in pots in soil to which root extract from 2-month old plants was added. Fungal counts of rhizosphere and non-rhizosphere soil showed a negative rhizosphere effect on both number of species and on fungal density. 7 amino acids and 3 sugers were detected by paper chromatography of the root extract. It was concluded that the root extract stimulated fungal growth in non-rhizosphere soil.

450.
MISHRA, M.M.; NEELAKANTAN, S. and KHANDELWAL, K.C. (1972) Effect of lindane and thimet on nitrification.

J. Res. HAU 2 : 283-285.

Neither lindane thimet affected the conversion of nitrite to nitrate at normal and at 10x normal dosage, and they had no effect on overall nitrification. Nitrite producing micro-organisms were affected by both pesticides but nitrate producing micro-organisms were not.

451.
MISHRA, M.M; NEELAKANTAN, S.; BHARDWAJ, S.K. and VYAS, S.R. (1973) Quantitative microbiological changes during decomposition of plant material in an alkali soil.

Zentbl. Bakt. Parasitkde Abt. II. 128 : 352-355.

The microbial changes during the decomposition of bajra stalks (Pennisetum glaucum) in an alkali soil were studied. Aspergillus spp., Cladosporium spp., Penicillium spp. and some Sphaeopsidales were the only permanent fungal flora. The fungi originating from the plant material did not survive in the soil for longer than six weeks. The most common bacteria during decomposition were Achromobacter spp., Arthrobacter spp., Bacillus spp., and Pseudomonas spp.

452.
MISHRA, M.M.; NEELAKANTAN, S.; KHANDELWAL, K.C.; BHARDWAJ, S.K. and VYAS, S.R. (1975) Margosa (neem) seed cake as an inhibitor of nitrification.

Soil Biol. Biochem. 7 : 183-184.

Margosa seed cake checked the conversion of ammonium to nitrite, presumably by its selective effects on ammonium-oxidizing bacteria. Better utilization of N by the plants is possible due to the slow release of nitrate during nitrification. Addition of seed cake at the rate of 0.2% C, which amounts to 8156 kg/ha, may assist the build up of soil organic matter, which is low in most Indian soils.

453.
MISHRA, M.M.; SINGH, C.P.; KAPOOR, K.K. and JAIN, M.K. (1979) Degradation of lignocellulosic material and humus formation by fungi.

Ann. Microbiol. 130 A : 481-486

Chaetomium globosum, Fusarium solani, Paecilomyces varioti and Penicillium chrysogenum produced little or no cellulase when grown on cellulose powder, in contrast to Aspergillus sp. and Trichoderma viride QM 9414. However, all produced cellulase when grown on wheat straw. All the cultures degraded cellulose and lignin to some extent, but P. varioti was the most efficient in cellulose and lignin degradation. There was an increase in humus-like substances after inoculation with fungi.

454.
MISHRA, M.M.; YADAV, K.S. and KAPOOR, K.K. (1981) Degradation of lignocellulose by mixed cultures of cellulolytic fungi and their competitive ability.

Zentbl. Bakt. Parasitkde. Abt. II, 136 : 603-608

The changes in C/N ratio, loss in weight, degradation of cellulose and lignin in wheat straw were studied after inoculation with mixed cultures of efficient cellulolytic fungi. The loss in weight per unit time was increased by inoculation with mixed cultures but no increase in cellulose degradation was observed. *Paecilomyces fusisporus* and *Trichoderma viride* showed the highest competitive ability for degradation and suppressed the growth of *Trichurus spiralis, Paecilomyces varioti, Penicillium chrysogenum* and *Sclerotium rolfsii.*

455.
MISHRA, R.P. and DAS, S.N. (1975) Attendant changes in the bacterial flora of rice rhizosphere as influenced by fertilizer treatment.

Pl. Soil 42 : 359-365.

The bacterial population of rhizosphere in both treated and untreated soils was significanty higher (91.5×10^6/g) than that found in non-rhizosphere soils (10.3×10^6/g). In rhizosphere soils there was no discernible trend in change of bacterial population with varying fertilizer levels. Lower numbers of aerobic heterotrophic bacteria were encountered in unfertilized non-rhizosphere soils and those receiving low levels of amendments (8.3×10^6/g) than in soils receiving high levels of NPK and farmyard manure (12.3×10^6/g). Bacteria requiring simple nutrition predominated in the rhizosphere of unamended soil (76.4 per cent) while those with complex requirement flourished in greater abundance (65.2 per cent) in rhizosphere of fertilized soils.

456.
MISHRA, R.R. (1967) Nature of rhizosphere fungal flora of certain plants.

Pl. Soil 27 : 162-166.

Fungal species derived from four different plants (Cynodon dactylon, Trichoderma amplexicaule, Abutilon indicum, Amaranthus spinosa) exhibited greater specificity in the rhizoplane region than in the non-rhizosphere.

457.
MISHRA, R.R. and KANAUJIA, R.S. (1972) Investigations into rhizosphere mycoflora. XIII. Effect of foliar application of certain plant extracts on Pennisetum typhoides f. burm Stapf & Hubb.

Israel J. agric. Res. 22 : 1-9.

In pot experiments the effect of foliar sprays of leaf extracts of Calotropis procera and Datura metal at concentrations up to 16% on the rhizosphere of pearl millet were studied. The fungal population decreased with increasing concentration of leaf extract, but 16% was slightly phytotoxic. Some fungal species reacted differently to the leaf extracts from the two plants.

458.
MISHRA, R.R. and KANAUJIA, R.S. (1973) Observations on soil fungistasis. III. Fungistasis in relation to soil depths, seasonal changes, soil amendments and physico-chemical characteristics of the soils.

Pl. Soil 38 : 321-330.

Generally, the fungistatic effect decreased with depth in the profile. It was higher in forest soil than in arable or grassland soil. It was highest during the rainy season when the fungal population was high. Physico-chemical characters seemed to affect it indirectly. The addition of various plant amendments, fungal spores, culture filtrates and mycelial extracts of certain fungi caused various levels of fungistasis in sterilized soil. Fungistasis thus seems to result from a combination of factors.

459.
MISHRA, R.R. and KANAUJIA, R.S. (1973) Studies on certain ecological aspects of soil fungi. II-Distribution of soil fungi in relation to cover vegetation and physico-chemical characters of the soil.

Anales Edafol. Agrobiol. 32 : 21-24.

The mycoflora of twenty one soil samples collected from different localities is described. Fungal population was generally higher in cultivated fields than in bare localities, like usar land and hillock tops. The longer the land was under a crop cover, the higher was the fungal population. The effects of the soil physico-chemical characteristics (organic matter, moisture, pH) on the soil mycoflora is discussed.

460.
MISHRA, A.K. and PRADOSH ROY (1979) A note on the growth of Thiobacillus ferrooxidans on solid medium.

J. Appl. Bacteriol. 47 : 289-292.

An iron oxidizing strain of Thiobacillus ferrooxidans has been grown on solid medium using purified agar, carrageenan (Type 1) and agarose. This strain produces isolated and transferable colonies after 7 day incubation. Growth (increase in viable cells) by the direct plating method has been followed in relation to iron oxidation. Acidity, agar concentration and phosphate influenced colony development on solid medium.

461.
MISHRA, A.K.; PAMPA ROY and ROY MAHAPATRA, S.S. (1983) Isolation of Thiobacillus ferrooxidans from various habitats and their growth pattern on solid medium.

Curr. Microbiol. 8 : 147-152.

Different strains of iron-oxidizing Thiobacillus ferrooxidans were grown and purified on solid medium containing Bapco agar, agarose, and carrageenan (Type 1). These strains produced easily countable isolated colonies that could be transferred after 7 days of incubation at 30°C. Increase in viable cell number in relation to growth and iron oxidation was studied by both microscopic count and direct plating method. Colony morphology of different strains growing on solid medium helped in differentiating the colony types.

462.
MISHRA, A.K.; PAMPA ROY; ROY MAHAPATRA, S.S. and CHANDRA, D. (1984) Low grade pyrites and their possible benefication by Thiobacillus ferrooxidans.

Proc. Indian natn. Sci. Acad. B50 : 519-524

An iron-oxidizing chemoautotroph, T. ferrooxidans was found to be quite efficient in oxidation of pyrite from pyriteferous shales. The oxidation was negligible in the absence of essential nutrients. The bacterium in association with the shales showed a great potential to recover copper from a copper ore tailings sample. Five different rock phosphate samples were also made vulnerable to bacterial attack with the help of shales. The leached liquor from the shales was utilized for preparation of chemicals such as ammonium sulfate and iron oxide mass. The latter being useful as a desulfurizing agent. In addition, other possible exploitations of such unutilized low grade pyrites have been discussed.

463.
MISHRA, D.K.; JHA, M. and KUMAR, H.D. (1985) Involvement of calcium in nitrogen fixation by Nostoc linckia.

Curr. Sci. 54 : 43-45.

This paper describes investigations on the nature of the Ca^{++} requirement for nitrogen fixation in the heterocystous cyanobacterium Nostoc linckia.

464.

MISHRA, R.R. and PANDEY, K.K. (1974) Studies on soil fungistasis: V. Effect of temperature, moisture content and incubation period.

Indian Phytopath. 27 : 475-479.

Fungistasis of soil samples from different localities was investigated using Curvularia lunata and Fusarium nivale as test organisms. Different fungistatic values for the two fungi could be attributed to differences in spore type.

465.
MISHRA, R.R. and PANDEY, K.K. (1981) Studies on soil fungistasis. Trop. Ecol. 22 : 211-218.

The fungistatic value of the soil samples collected from the rhizosphere region of Cajanus cajan and Lathyrus sativus had been assessed by indirect, direct and soil emanation agar methods. The fungistatic property of the soil samples collected at different periods varied for all the test fungi.

466.
MISHRA, R.R. and SHARMA, G.D. (1977) Ecology of soil fungi: population variation in relation to varying cover vegetation and soil factors. Sydowia Ann. Mycol. 30 : 134-140.

The role of cover vegetation on the distribution of soil microflora of three ecosystems in northeast India was studied. The ecosystems were three forests (Tectona sp.; Shorea robusta; Terminalia sp. with Phyllanthus emblica), a grassland (Saccharum spontaneum and Themeda sp.) and barren land (a hot spring bed and adjacent land). The effect of physicochemical characters of soil (moisture, pH, temperature, conductivity, organic C, available P and available K) on microflora was also investigated. In the fungistatic study the correlation between occurrence of species and soil inhibition of spore germination was not significant.

467.
MISHRA, R.R.; SHARMA, G.D. and GATPHOH, A.R. (1980) Mycorrhizas in the ferns of north eastern India. Proc. Indian natn. Sci. Acad. 46B : 546-551.

All of the 40 fern species examined possessed vesicular-arbuscular mycorrhiza. The endophytes belonged to the genera Glomus, Gigaspora, Acaulospora and Sclerocystis. Sclerocystis was associated with all except three of the fern species. Infection levels ranged from 63.3 to 92.2%. Different fern species exhibited different stages of fungal invasion ranging from hyphae, arbuscules and vesicles or a combination of all the three.

468.
MISHRA, S.G. (1981) Jiwanu Urvarak. Indian Council of Agriculture Research, New Delhi. p 109.

The book stresses the role of biofertilizers particularly azotobacter and rhizobium in improving nitrogen fertility and phosphate bacterium improving the availability of phosphorus. The rolo of algalization in improving yields of paddy has also been highlighted.

469.
MISRA, A. (1983) Ultrastructural aspects of nodulation and legume Rhizobium symbiotic responses. Proc. Natl. Symp. Adv. Front. Pl. Sci. Jodhpur. 9 : 100-102.

The symbiotic event in legume root is initiated in the rhizosphere with selective stimulation of Rhizobial strains, which get attached to root hair mucilage. Infection of Rhizobium starts with root tip curling and formation of infection thread by invagination and ingrowth in root hair that reaches the cortical region.

470.
MISRA, P.C (1967) Torula terrestris sp. from soil.

Can. J. Bot. 45 : 367-369.

The hyphomycete was first isolated from soil of a Shorea robusta forest.

471.
MISRA, V.K. and ABHISWAR SEN (1957) Effect of low and high nitrogen materials on **nodulation** of pea in Delhi soil.

J. Indian Soc. Soil Sci. 5 : 247-251

Pea was raised in Delhi soil with treatments of mahua cake, ammonium nitrate, urea and mixtures of mahua cake with ammonium nitrate and urea at the rate of 100 lb. nitrogen per acre.

Nodules on pea treated with mahua cake were small, whitish and distributed all over the lateral roots. Nodules on pea with other treatments were larger, with two or four lobes and distributed on tap along with lateral roots.

Treatment of mahua cake had an injurious effect on the weight of nodules but did not affect the numbers of nodules on pea plants as compared with those produced under other treatments.

Number of nodules on pea had no relation with the final yield of the crop. Inspite of the differences in the characteristics of the nodules under different treatments, the yields of whole plants and the grains were significantly related to the weight of the nodules found after growth for ten weeks.

472.
MITRA, A.K. (1950) Two **new algae** from Indian soils. Ann. Bot. 14 : 457-464.

A new monotypic genus Chloranomala palmelloides and a new species, Chlorogloea fritschu, are described.

473.
MITRA, S.P. and PRAKASH, D. (1955) Fixation of atmospheric nitrogen under sterile conditions on kaolinite with straw and cow dung and the effect of different phosphates on the effi-

ciency of fixation. Proc. nat. Acad. Sci. India 24 A : 187-195.

Nitrogen fixation apparently proceeded, both in light and darkness, through oxidation of the organic matter and was much promoted by various phosphates.

474.
MODI, V.V. and PATEL, R.N. (1968) Salicylate formation from naphthalene by *Pseudomonas aeruginosa.*

Appl. Microbiol. 16 : 172-173.

P. aeruginosa isolated from oil-rich fields utilized diesel oil, kerosene, paraffin, petroleum waste, or naphthalene as a sole source of C in a culture medium of mineral salts. Salicylate was produced as a metabolite, but B or Cu inhibited salicylate formation.

475.
MOHAN SINGH (1984) Associative nitrogen fixation by sugarcane.

J. Indian Soc. Soil Sci. 32 : 175 177.

Among the several isolates of *Azotobacter chrococcum,* a most efficient strain was found to fix 16 mg N/g glucose consumed. This strain when inoculated in sugarcane plant produced significantly higher dry matter as well as nitrogen content in comparison to the control. The effectiveness of *A. chrococcum* inoculation as nitrogen fixing bacteria in sugarcane has been clearly demonstrated in this study.

476.
MUKERJI, K.G. and SUBBA RAO, N.S. (1982) Plant surface microflora and plant nutrition. In advances in Agricultural Microbiology Ed. by N.S. Subba Rao. Oxford & IBH Publishing Co. New Delhi, 111-138

The article gives a comprehensive account of plant surface microflora such as those of rhizosphere, rhizoplane and mycorrhiza, phyllosphere, phylloplane and leaf nodulation and caulosphere and cauloplane in relation plant nutrition.

477.
MUKERJI, K.G.; BHATTACHARJEE, M. and TEWARI, J.P. (1983) New species of vesicular arbuscular mycorrhizal fungi. Trans. brit. Mycol. Soc. 81 : 641-643.

Glomus delhiense and G. multisubstensum are newly described from Indian soils. In pot tests the former formed VAM with Trigonella foenum-graecum and the latter with maize.

478.
MUKERJI, S.K. (1937) Changes in soil nitrogen after the addition of fresh cow dung to soil. Proc. nat. Acad. Sci. India 7 : 158-164.

Atmospheric N was fixed when cow dung was mixed with soil in dishes, field plots and sterile vessels, expecially in sunlight. Molasses seemed to stimulate the oxidation of manure and increase the amount of N fixed. Applying cow dung and cellulosic material to the field is considered superior to composting. The fixation process is considered to be partly photochemical, catalytic and thermal as well as bacterial.

479.
MUKERJI, S.K. (1938) Denitrification in sunlight and its retardation. Part IV. Allahabad Univ. Stud. 14 : 71-76.

When the C content of a soil is high and the N low, there will be less oxidation of the soil N and less chance for the formation and decomposition of ammonium nitrate and less N loss than when the C content is low and the N content high. These conclusions justify the application of organic manures for preserving the N in soils, expecially in tropical countries where the N compounds require more protection than in temperate climates.

480.
MUKERJI, S.K. (1938) Photonitrification and the influence of temperature on the nitrate formation in the soil. Allahabad

<u>Univ. Stud.</u> 14 : 63-69.

The optimum temperature for nitrate accumulation in soil in the presence of the soil microflora is between 25°-30° and not above this temperature. Slight nitrate formation was observed at 60°.

481.
MUKHERJEE, S.K. and NANDI, P.N. (1955) Studies on the antagonistic actinomycetes from the soils of West Bengal. <u>Indian Phytopath.</u> 8 : 64-71.

Of 170 strains isolated, 40% were inhibitory to grampositive bacteria, 21% to gram negative bacteria and 32% to higher fungi.

482.
MURTHY, K.S.; NEHRU BABU and ELLAIAH, P. (1973) Isolation of a new <u>Streptomyces</u> from a soil sample of Visakhapatnam. <u>Hindustan Antibiot. Bull.</u> 15 : 156-157

A new <u>Streptomyces</u> sp. <u>S. Visakhaensis</u> has been isolated from a soil sample collected near the port bed area of Visakhapatnam. Its morphological, cultural and physiological characteristics are reported in detail.

483.
MUTHUSWAMY, P.; RAJU, G.S.N. and KRISHNAMOORTHY, K.K. (1975) Mineralization of urea coated with nitrification inhibitors. <u>J. Indian Soc. Soil Sci.</u> 23 : 332-335

A pot experiment with red, black and alluvial soils using ragi (<u>Eleusine coracana</u> Gaerin.) as a test crop was conducted with urea treated with sulphathiazol, neem oil, neem cake, mahua cake, coal tar extract and with two forms of urea viz., lac coated urea and CDU. The urea was applied at two levels viz., @ 50 and 100 kg N/ha. The results of soil analysis at weekly intervals revealed that the exchangeable NH_4 + and

NO_3 - decreased at the crop growth advanced. The N applied @ 100 kg/ha recorded the maximum NH_4 and NO_3 - in the soil in all the periods of analysis. The maximum yield of ragi grain was recorded in the treatments which received urea treated with coal tar and mahua cake extracts, whereas maximum straw yield was recorded in lac coated urea along with coal tar and mahua cake extracts. Black soil significantly registered higher yield than red, followed by alluvial soils. Both grain and straw yields were significantly higher in 100 kg N/ha rate.

484.
NAGARAJ RAO, H.S. (1974) Response of groundnut variety TMV_2 to Rhizobial inoculation. Proc. Indian natn. Sci. Acad. 30B : 650-651.

Four different strains of Rhizobium GN-2 , 402-A, 402-B and 403-C: were tested for their efficiency against the groundnut crop variety, TMV_2 in a pot culture trial at the Division of Microbiology, Indian Agricultural Research Institute, New Delhi, and in a field trial at the Regional Research Station, I.A.R.I., Rajendranagar, Hyderabad-30. In the field trial the GN-2 strain was also used as a peat based culture (GNP-2). Strain 402-B was found to be most effective in significantly increasing the yields, nitrogen content and total nitrogen uptake by groundnut.

485.
NAINAWATEE, H.S. SANTOSH RATHEE and WAGLE, D.S. (1978) Amideamino acid synthetase of Phaseolus aureus Rhizobium phaseoli symbiotic system. Pp.296. In S.P. Sen, Y.P. Abrol and S.K. Sinha (eds.), Nitrogen Assimilation and Crop Productivity. Associated Publishing Company, New Delhi.

Glutamine synthetase and aspargine synthetase activities have been studied for 10 weeks in nodule, root and leaf tissues of the Phaselous aureus Rhizobium Phaseoli Symbiotic System. Both the enzyme activities were high in nodules as compared to root and leaves. The ratio of glutamine synthetase and aspargine synthetase activities were almost constant in nodules but fluctuated both in roots and leaves. A route of ammonia assimilation has been proposed.

486.
NAIR, G.A. (1978) Some aspects of the population characteristics of the soil isopod, <u>Porcellio laevis</u> (Latreille), in Delhi Region.

<u>Zool. Anzeiger</u> 201 : 86-96

In <u>Porcellio laevis</u>, the embryo standing crop figures for the two stations studied showed a steady decline from March/April to June/July, while in August a steep rise in standing crop was observed following the onset of monsoon. The total embryo production in 1971 was much higher than that of 1970 in both stations because of the early onset of the monsoon. Following the breeding period, there was an increased mortality among the adults. Factors such as high temperature and low humidity of the soil also increased the overall mortality. It was also noted that the monsoon recruits of 1969 and spring recruits of 1971 showed greater survival than the spring and monsoon recruits of 1970.

487.
NAIR, K.S.; KRISHNA MENON, K.M. and MARIAKULANDAI, A. (1957) Effect of 2,4-D on soil micro organism.

<u>Madras agric. J.</u> 44 : 667-668

The analysis of the data reveal that the total bacterial population is not significantly affected due to spraying during the first 15 days. A marked rise in numbers is found from 15 to 30 days, increase being significantly higher than the control. The effect of 2,4-D persists upto 45 days and thereafter no important effect is noticed, with may be due to the almost complete breakdown of the 2,4-D in the soils, by that period. The amount of carbondioxide evolved from the soils followed a similar trend as in the case of the total bacterial counts. Favourable effect on the growth of <u>Azotobacter</u> was also indicated. 2,4-D suppressed the growth of denitrifying and nitrate reducing bacteria. Cellulose decomposing bacteria remained unaffected throughout the entire period, by the application of 2,4-D.

488.
NAIR, S.K. and SUBBA RAO, N.S. (1977) Distribution and activity of phosphate solubilising microorganisms in the rhizosphere of coconut and Cacao under mixed cropping.

J. Plantation Crops, 5 (2) : 67-70.

Two efficient phosphate solubilising microorganisms, Pseudomonas sp. and Asperqillus niger, were isolated from the rhizosphere of coconut and cacao. They were capable of solubilising 49.0% and 49.7% of the phosphate in tricalcium phosphate. The incidence of phosphate solubilising microorganisms and available phosphorus in different rhizosphere soils were directly related.

489.
NAMBIAR, P.T.C.; RAVISHANKAR, H.N. and DART, P.J. (1983) Effect of Rhizobium numbers on nodulation and dinitrogen fixation in groundnut. Expl. Agric. 19 : 243-250.

Glasshouse studies have shown that nodulation and dinitrogen fixation in groundnuts grown in a sand/vermiculite mixture increased with an increase in Rhizobium numbers inoculated onto each seed. Cultivars Robut 33-1 and TMV 2 required 10^6 rhizobia per seed for maximum nodulation and dinitrogen fixation, while cultivar Kadiri 71-1 required larger numbers.

490.
NAMDEO, K.N. and DUBE, J.N. (1973) Proteinase enzyme as influenced by urea and herbicides applied to grassland oxisol.

Indian J. exp. Biol. 11 : 177-119.

Urea, dalapon and paraquat each increased urease activity and decreased proteinase activity. The combination of urea plus paraquat increased proteinase and decreased numbers of ureolytic actinomycetes.

491.
NAMDEO, S.L.; DUBE, J.N. and VAISHYA, U.K. (1974) Relataive performance of Rhizobium japonicum serotypes in soil inoculum. Proc. Indian natn. Sci. Acad. 40B : 488-490.

Field soil subjected to repeated inoculations with the effective commercial inoculants over four years was fortified with a range of nutrients; gluconate, pantothenate and biotin, cobalt and molybdenum and was further protected with mineral oil. Three dominant serotypes were recognised in the fortified and control samples. These two types of soils were used as incoula at the standard rate of 200 g/ha. Field and green house tests were conducted, and nodule reisolates were checked for the dominant serotypes. The distribution and dominance of Rhizobium japonicum population in the soil inocula are discussed.

492.
NANDI, A.S.; GUPTA, B.S. and SEN, S.P. (1983) Utility of phyllosphere N_2-fixing microorganism in the improvement of growth of some vegetables. J. Hort. Sci. 58 : 547-554.

Three N_2-fixing strains of Klebsiella, namely, $KUPBR_2$, KUP_4 and KUP_6D, were sprayed on cauliflower, cabbage, tomato, pumpkin, eggplant, potato, onion, radish, spinatch and Trichosanthes dioica leaves 3 times at fortnightly intervals. Significant increased were observed in the growth, yield and N content of all plants. Yields, although always less than those obtained using urea, were doubled for several species compared with the non-treated controls.

493.
NANDI, P. and SINHA, N. (1974) Mass scale production of Rhizobium culture. Proc. Indian natn. Sci. Acad. 40B : 479-481.

Experiments were conducted for mass scale production of Rhizobium inoculant in different cheap carbon sources and in different concentrations of yeast extracts in the fermentation medium.

Molasses could serve as a superior carbon source for rhizobila growth. Liquid malt extract could be ranked in the list and third one was the medium having sucrose and mannitol in combination.

Neutral Baker's yeast extract solution 200 ml/l or Oxoid yeast extract 500 mg/l and Difco 250 mg/l served as a very suitable substrate for growth. The modified composition of medium comprised 10 g sucrose and 5 g mannitol and 200 ml yeast extract solution/1 and served as a very suitable composition for rapid cell multiplication.

494.
NARAIN, P. and DATTA, N.P. (1973) Influence of nitrification inhibitors and some insecticides on nitrification of urea under aerobic and water-logged conditions.

Agrochimica 18 : 79-89.

Five typical Indian soils, alluvial soil, saline alkali soil, black-cotton soil, laterite soil and red soil were supplied with 200 ppm-N as urea, and were incubated with 5 ppm N-Serve or 2-amino-4-chloro-6-methyl pyrimidine (AM), or with insecticides. Under aerobic conditions N-Serve and AM inhibited nitrification in red, black-cotton and alluvial soils; in laterite soil, only N-Serve was effective and in saline alkali soil neither had any effect. Insecticides had no effect on nitrification and in water-logged conditions neither the inhibitors nor the insecticides had any effect.

495.
NATARAJAN, S. (1956) Carbon nitrogen metabolism of soil fungi. Proc. Indian Acad. Sci. 44 : 289-310.

The total N content of the fungal mat and of the cultures decreased during incubation. In *Fusarium udum* culture the pH increased more rapidly at low sucrose levels than with *F. vasinfectum*. In both fungi, rates of sucrose deletion and N accumulation followed an exponential law.

496.
NATARAJAN, S. (1958) Carbon-nitrogen metabolism of soil fungi. IV - Influence of carbon sources. Proc. Indian Acad. Sci. 47B : 338-344.

Irrespective of the N source ($NaNO_3$, NH_4Cl etc.) the

efficiency of C source on mycelial weight of *Fusarium vasinfectum* and *F. udum* and N contents in the mat was in the order : fructose > sucrose > maltose > glucose > mannose > lactose > mannitol > galactose > sorbitol, indicating that N metabolism is closely associated with differences in the utilization by the fungus of the energy material available.

497.
NATARAJAN, S. (1958). Carbon-nitrogen metabolism of soil fungi. V-Role of cyanide, nitrite and sulphur. *Proc. Indian Acad. Sci.* 48 : 28-32.

Fusarium vasinfectum and *F. udum* in nutrient solution containing $NaNO_3$ and different monosaccharides or disaccharides reduced NO_3 to NO_2 and NH_3 irrespective of the carbon source. The fact that high concentration of added KNO_3 proved to be toxic indicates that the reduction of NO_2 to NH_3 is more rapid than that of NO_3 to NO_2. Treatment with KCN at a concentration of 4 mg/50 mg did not affect fungal growth an NO_2 production, suggesting the presence of a KCN insensitive nitrate reductase. Treatment with S improved fungal growth and was accompanied by the formation of some H_2S and absence of NO_2, indicating that it acted as a H acceptor and an oxidizing agent towards H_2S.

498.
NAYAK, D.N. and RAJARAMAMOHAN RAO, V. (1977) Nitrogen fixation by *Spirillum* sp. from rice roots.

Arch. Microbiol. 115 : 359-360.

In culture medium, N fixation by *Spirillum* isolated from the roots of irrigated rice to which ammonium sulphate at 0-100 kg/ha N had been applied was greater after application of 20-40 kg N/ha to the rice than it was after application of 60-100 kg /ha N. The N-fixing efficiency of *Spirillum* varied with plant age, irrespective of the rate of application of N.

499.
NAYAK, D.N. and RAO, V.R. (1980) Pesticides and heterotrophic

nitrogen fixation in paddy soils.

Soil Biol. Biochem. 12 : 1-4

The effect of three pesticides, benomyl, carbofuran and gamma-BHC at 5 parts /10^6, rates equivalent to recommended field levels, on the heterotrophic N_2 fixation in five air-dried, cellulose-amended, submerged, tropical soils were investigated employing 15N tracer technique under laboratory conditions, laterite and two acid sulphate soil resulted in significant increases in N_2 fixation. Carbofuran, a methylcarbamate insecticide, had a stimulatory effect on N_2 fixation in alluvial laterite and acid saline soils. Gamma-BHC, a chlorinated hydrocarbon insecticide, stimulated N_2 fixation in alluvial and acid solphate pokkali soils, while considerable inhibition of N_2 fixation was evident in other soils. Results showed differential responses of specific groups of N_2-fixing organisms to the pesticides depending on the soil type.

500.
NAYAK, D.N. and RAO, V.R. (1982) Pesticides and nitrogen fixation in paddy soil. Soil Biol. Biochem. 14 : 207-210.

While the application of diazinon to a submerged paddy soil under laboratory conditions had no marked effect on 15 N_2 incorporation and C_2H_2 reductions benomyl, carbofuran, parathion, nitrofen and Y-HCN at concentrations close to recommended field application rates significantly stimulated N_2 fixation. Synergistic stimulantory effects of the pesticides on N_2 fixation were evident, particularly in combinations of carbofuran with benomyl, nitrogen and Y-HCN. On the contrary, diazinon slightly retarded the stimulatory effect of benomyl and carbofuran. The differential effects of pesticides on N2 fixation could be attributed partly to fluctuations in the population of certain groups of N_2 fixers in submerged soil.

501.
NAYAK, D.N.; CHARYULU, P.B.B.N. and RAJARAMAMOHAN RAO, V. (1981) 15N_2 incorporation and acetylene reduction by Azospirillum isolated from rice roots and soils.

Pl. Soil 61 : 429-436

Nitrogen fixation by strains of Azospirilum isolated from seven rice soils with widely differing physico-chemical properties, and from seven rice cultivars, varied markedly, as measured by $15N_2$ incorporation and C_2H_2 reduction.

502.
NAYAK, N.D. and RAJARAMAMOHAN RAO, V. (1981) The influence of alternate flooded and nonflooded conditions on nitrogen fixation (C_2H_2 reduction) in paddy soils. Soil Sci. 131 : 26-29.

In two paddy soils amended with cellulose, nitrogenase activity was higher under submerged than under nonflooded conditions. Nitrogenase activity increased several times following a shift from nonflooded to flooded conditions, and decreased when the flooded soil was returned to the nonflooded condition by draining. Soil submergence favoured nitrogen fixation irrespective of the length of the pretreatment under nonflooded conditions.

503.
NAYYAR, V.K.; RANDHAWA, N.S. and CHOPRA, S.L. (1970) Effect of simazine on nitrification and microbial population in a sandy-loam soil.

Indian J. agric. Sci. 40 : 445-451.

Simazine at up to 2.5 ppm did not affect transformation of NH_4-N to NO_3-N in a sandyloam soil. Concentrations up to 10 ppm did not affect oxidation of NH_4-N to NO_3-N, but markedly inhibited transformation of NO_2 to NO_3 during the first half of incubation. Simazine decreased the size of the bacterial population, but slightly increased the fungal population.

504.
NEELAKANTAN, S. and PAHWA, M.R. (1977) Note on the distribution of micro-organisms in normal, saline and alkali soils. Curr. Agric. 1(1-2) : 77-78.

The normal, saline (EC_2 6.10 mmhos/cm) and alkali (pH_2 10) soils contained bacteria 96.00, 11.30 and 2.33, actinomycetes 11.6, 0.4 and 0.04 fungi 0.21, 0.045 and 0.00003 X 10^5/g soil, respectively.

505.
NIGAM S.S.; RANGANATHAN, S.K. and GUPTA, S.R.S. (1960) Microbial degradation of cotton cellulose in soil. J. Sci. ind. Res. 19 C : 20-24.

Bacteria and actinomycetes are largely present during the initial stages of degradation of cotton cellulose buried in soil to cause considerable decay; fungi appear later. The optimum moisture content of soil for degradation of cotton cellulose is 25-30%.

506.
NORRIS, B.V. and SUBRAHMANYAN, V. (1929) Soil Actinomycetes. Agric. J. India 24 : 232-234.

A strach agar medium containing a minimum amount of nitrogen with the necessary minerals was best suited as the count medium. The counts of actinomycetes thus obtained for Indian soils varied from nil to about 4 millions per gramme of soil. The climatic conditions, rainfall and physical texture of the soil did not, by themselves, influence the numbers to any appreciable extent.

507.
OBLISAMI, G. (1974) Studies on the *Rhizobium* and nodulation pattern in a forage legume, *Clitoria ternatea* Linn. Proc. Indian natn. Sci. Acad. 40B : 618-623.

Rhizobium sp. isolated from an important, drought-resistant forage legume, *Clitoria ternatea,* was found to grow better in yeast extract mannitol agar and carrot extract agar. It exhibited good growth with the carbon sources like xylose, glucose, sucrose and mannitol, while fructose, galactose, lactose, faffinose, glycerol and sorbitol caused poor growth. The *Rhizobium* from *C. ternatea* developed nodules on cowpea and soyabean cross-inoculation group plants and vice-versa. More numbei of nodules were formed on the lateral roots than the tap roots in all the plants. The plants belonging to alfalfa, clover, bean and pea cross-inoculation groups were not nodulated and vice-versa.

508.
OBLISAMI, G. and RANGASWAMI, G. (1967) Qualitative and quantitative variations in the antagonistic property of different strains of four species of soil actinomycetes.

Mysore J. agric. Sci. 1 : 274-283.

Strains of Streptomyces lavendulae, S. griseus, S. albus, and S. hygroscopicus were obtained, at 0-3, 3-6 and 6-12 inch depths, from seven soils, representing the black, alluvial, red and laterite soils of south India. Different strains of the same species had significantly different antagonistic properties, but there was no definite correlation between antagonistic properties and soil type. Strains from different layers inhibit differentially against various bacteria and fungi. Most strains of S. lavendulae and S. hygroscopicus are antibacterial and antifungal, S. griseus and S. albus strains are mainly antibacterial.

509.
OBLISAMI, G.; BALARAMAN, K.; VENKATARAMAN, C.V. and RANGASWAMI, G. (1973) Effect of three granular insecticides on the growth of Rhizobium from redgram.

Madras agric. J. 60 : 462-464.

Granular formulations of endrin, carbofuran and disulfoton inhibited the growth in vitro of the redgram rhizobium; endrin was most harmful and disolfoton least harmful. The inhibitory effect increased with increasing concentration of active ingredients in the case of endrin and carbofuran, but only to a small extent with disolfoton.

510.
PADMAVATHY, K. and RAMARAO, P. (1970) Two ascomycetes

new to Indian soils.

Indian Phytopath. 23 : 589-590.

Arachniotus dankaliensis was isolated from a cabbage field soil and Thielavia terricola var. minor from a costal soil.

511.
PAHWA, K. and DOGRA, R.C. (1981) H_2-recycling system in mungbean Rhizobium in relation to N_2 - fixation.

Arch. Microbiol. 129 : 380-383.

Thirty isolates of mungbean Rhizobium were tested for the presence of a H_2-recycling system. The six isolates which reduced triphenyltetrazolium chloride rapidly in culture at early stages of growth recycled hydrogen in vivo and in vitro. The nitrogen fixing effeciency of hydrogenase positive and hydrogenase negative isolates and Hup^- mutants was compared in greenhouse experiments. There was a 13-56% increase in dry matter and a 21-46% increase in total introgen in mungbean plants inoculated with H_2-recycling isolates in comparison with plants inoculated with non-recycling isolates. There were reductions in dry matter and total nitrogen contents of plants incoulated with Hup- mutants, in comparison with plants inoculated with a wild type strain, of 19-22% and 20-26%, respectively.

512.
PAHWA, M.R. and PATIL, B.D. (1984) Response of forage crops to Azospirillum brasilense inoculation. Golden Jubilee Seminar on Soil Resources and Productivity Management, Indian Society of Soil Science, New Delhi 3 : 55-56.

Field studies were laidout in different soil types of Bundelkhand region to test the effect of Azospirillum brasilense inoculation with and without urea on the yield of forage crops. Simple seed inoculation registered increased green fodder yield to the tune of 18.6, 32.3, 30.9, 41.4 and 38.5 per cent in teosinte, maize, bajra, oats and barley, respectively. The corresponding rise in dry matter production was also observed. The results showed 15-20 kg N/ha saving of inorganic nitrogen in the form of urea with the inoculation in cereal forages. Significant responses were noticed in kabar (heary textured) soil as compared to light textrued ones.

513.
PAL, S. and CHHONKAR, P.K. (1979) Thermal sensitivity and kinetic properties of soil urease.

J. Indian Sc. Soil Sci. 27 : 43-47

Thermal sensitivity, pH optima and reacton rate constants of urease in soils of the Indian Agricultural Research Institute farm were studied. Influence of different soil characteristics on the enzyme kinetics was also evaluated. Soil urease activity was not adversely affected between 40 and 70°C but at 80°C and higher temperatures the enzymes was completely inactivated. In buffered systems peak urease activity was observed at pH 6.6 and 9.25. Michaelis constant (km) for the enzyme varied in different soils and out of the various soil characteristics studied, it was significantly and positively correlated with soluble salt content only.

514.
PALAN, I.V. and MARATHE, K.V. (1978) Observations on the effect of temperature on soil algae.

J. Indian Bot. Soc. 57 (Suppl.) : 3

From investigations on four soil samples collected from different localities, it was found that blue-green algae and unicelluar green algae were very resistant to high temperature. Low temperature had no effect on the algal flora. The maximum temperature tolerated by blue-green algae was 95-136°C, and by green algae, 55-135°C. Diatoms tolerated temperatures of up to 90°C.

515.
PALANIAPPAN, R. and RAJ, D. (1973) Studies on denitrification in South Indian soils.

Madras agric. J. 60 : 86-89.

Representative soils were incubated under waterlogged conditions with KNO_3 or NH_4NO_3 and glucose. Loss of NO_3- N increased with time of incubation and was similar for both N sources. The highest and lowest losses were from black soils and lateritic soils, respectively, with red soils showing intermediate values.

516.
PALANIAPPAN, S.P. and BALASUBRAMANIAN, A. (1983) Effect of two pesticides on oxidative metabolism of cowpea bacteroids. Curr. Sci. 52 : 674-675.

Results show that furadan applied to soil had no effect on the oxidation of pyruvate, citrate, succinate or fumarate, but significantly increased the oxidation of acetate.

517.
PALARPWAR, M.Y. (1983) Use of Azotobacter in wheat.

Madras agric. J. 70 : 835-836.

The grain yield of wheat increased by 16 to 33 per cent due to inoculation. Inoculation of seed with Azotobacter also improved the germination by 30 per cent.

518.
PANDEY, B.N. and DWIVEDI, R.S. (1979) Role of iron and zinc on the colonization of Aspergillus terreus and Penicillium citrinum in rhizosphere of Solanum melongena L. Acta Bot. Indica 7 : 96-97.

Zn and Fe at concentrations of up to 100 ppm promoted colonization of the rhizosphere of S. melongena by Aspergillus terreus and Penicillium citrinum. The maximum number of fungi/g was recorded at 25 ppm concentration. Colonization was greater in rhizosphere cultures (prepared by sowing seeds of S. melongena in coarse sand) than in non-rhizosphere cultures.

519.
PANDEY, S.N. (1969) Effect of nitrogen, phosphorus, potassium and molybdenum on nodule formation.

Indian J. Agron. 14 : 205-207.

In pot experiments with *Trifolium alexandrinum* and *Cicer arietinum* medium rates on N combined with a high rate of P and a low rate of Mo were optimum for nodulation, which was apparently promoted by increased secretion of IAA in treated plants.

520.
PANDEYA, J.P.N. and SAKSENA, S.B. (1978) A new species of Polyschema from Indian soil.

Mycologia 70 : 876-879.

Polyschema sagari Pandeya and Saksena, sp. nov. is described from the forest litter of red surface soil in the Patharia forest of Sagar.

521.
PANDIAN, M.R. and BALASUBRAMANIAN, M. (1978) Effect of phorate on the rhizosphere microflora of paddy in four soil types.

Il Riso 27 : 211-216.

In a pot experiment, phorate 10G applied to four soil types increased the bacterial and actinomycete populations in the rhizosphere by 18.5 and 12.48% respectively, but decreased the fungal population by 26.2%. The number of bacteria, fungi and actinomycetes in the soils decreased in the order clay soil, black cotton soil, red soil, sandy loam.

522.
PANT, S.D. and ISWARAN, V. (1970) Survival of groundnut *Rhizobium* in Indian soils.

Mysore J. agric. Sci. 4 : 19-25.

A *Rhizobium* sp. from groundnut introduced into 7 different soils decreased in numbers with time; the survival period ranged from 55 weeks in Coimbatore black soil under sterilized and constant moisture conditions to 16 weeks in Coimbatore red soil under unsterilized and undisturbed conditions. No correlation was found between the longevity of *Rhizobium* and different soil constituents.

523.
PARICHA, P.C. and KAR, M. (1983) Improving symbiotic nitrogen fixation and productivity in black gram (*Vigna mungo* L.)

Curr. Sci. 52 : 1143-1145.

The present study was undertaken to know the response of Mo seed treatment and to relate the benefits in terms of applied nitrogen to an acid laterite soil of pH 5.4. The nodule number and dry weight of nodules increased due to M_o. Simulation in N concentration due to M_o alone was equivalent to the highest level of N application (45 kg/ha).

524.
PARVATHI, K.; VENKATESWARLU, K. and RAO, A.S. (1984) Occurrence of VA-mycorrhizas on different legumes in a laterate soil.

Curr. Sci. 53 : 1254-1255.

Results show that mycorrhizal fungi colonize effectively in the roots of a variety of leguminous plants growing in the infertile lateritic soil. *Glomus mosseae* was found to be the most widespread VAM fungus in this soil type.

525.
PATHAK, A.B. and JAIN, S.L. (1965) Effect of alkali salts : (ii) on nitrification. *J. Soil Wat. Conserv. India,* 13 : 30-32.

Effect of alkali salts on nitrifying organisms was studied both in liquid and soil media. Irrespective of salts and their concentrations, soil culture showed better nitrification than the liquid. Calcium carbonate was beneficial even up to higher concentration whereas the other salts proved toxic beyond a certain limit.

526.
PATHAK, A.N. (1953) Nitrification in Kanpur soils. Allahabad Fmr. 27 : 81-86.

Nitrification was higher in April and October-November. Unmanured soil had greater nitrifying capacity than manured soil, and nitrification was greater in clay soils than in silt and sand.

527.
PATHAK, A.N. (1954) Effect of manurial treatments on microbiological population of the soil. J. Soil Wat. Conserv. India 2 : 69-75.

Counts by the plate method were made on incubated soil samples taken from unfertilized pots and from pots treated with ammonium sulphate, compost or FYM. The greatest increase in the numbers of fungi and actinomycetes occurred after treatment with organic manures and in those of bacteria after ammonium sulphate. A highly significant correlation was observed between microbial population and CO_2 evolution in the soil.

528.
PATHAK, A.N. and JAIN, S.L. (1965) Effect of alkali salts : II. On nitrification.

J. Soil Wat. Conserv. India 13 : 30-32.

In laboratory tests involving increasing concentrations of various salts in inoculated liquid media and soil, nitrification was promoted by $CaCO_3$ even in high concentration (up to 1.6%) and by $CaSO_4$ and $MgSO_4$ up to a certain concentration and was inhibited even at low concentrations (given in brackets) of NaCl (0.02%), Na_2SO_4 (1%), Na_2CO_3 (0.2%) and $NaHCO_3$ (0.2%).

529.
PATHAK, A.N. and LEHRI, L.K. (1959) Studies on termite nests I. chemical, physical and biological characteristics of a termitarium in relation to its surroundings.

J. Indian Soc. Soil Sci. 7 : 87-90.

A study of the soil from a termite nest showed higher values for oven-dry moisture, loss on ignition, CaO, MgO, P_2O_5, and carbon and nitrogen as compared to the surrounding soil. Termite nest soil was rich in finer particles which were in the form of aggregates of bigger size. pH value of termite nest soil was 4.5 as against 8.1 of the control. When the pH value was adjusted with lime, termite nest soil showed greater biological activity than the control soil.

530.
PATHAK, A.N. and SHRIKHANDE, J.G. (1952) Effect of different mechanical fractions on the efficiency of nitrifying organisms. Curr. Sci. 21-13.

The formation of NO_3 was greater in presence of clay than of the coarser fractions and all three mechanical fractions separately encouraged nitrification to a higher degree in samples from unmanured than from manured plots.

531.
PATHAK, A.N. and SHRIKHANDE, J.G. (1953) Optimum temperature for nitrification and nitrogen fixation.

J. Indian Soc. Soil Sci. 1 : 131-136

The physiological activity of an Azotobacter culture isolated from Kanpur soil was studied both in pure and crude culture in liquid and sand media. $40^{o}C$ was found to be the optimum temperature for nitrogen fixation. Greater nitrogen fixation occurred in sand medium than in the liquid medium. Nitrate formation was found to be maximum at $36^{o}C$. There was practically no difference in the amount of nitrite formed in liquid and sand media. The optimum temperature of nitrate formation by nitrifying bacteria in pure and crude culture in liquid media was found to be 30°C. In sand medium the optimum was between 36°C and 38°C. In the mixed culture isolated from soil the amount of nitrate formed was adversly affected.

532.
PATHAK, A.N., HARI SHANKAR and AWASTHI, K.S. (1961) Effect of some pesticides on available nutrients and soil microflora.

J. Indian Soc. Soil. Sci. 9 : 197-200.

Effect of D.D.T., chlordane, benzene hexachloride and aldrin on soil fertility was studied. None of the insecticides tested was toxic to total bacteria and fungal population of the soil. Nitrogen fixing and nitrifying capacity of the soil were also not injured to the least. The insecticides tested, for the doses, rather stimulated the growth and activity of the soil organisms. They also increased the availability of phosphorus in the soil.

533.
PATHAK, A.N.; SHRIKHANDE, J.G. and MUKERJI, S.K. (1951) Nitrogen fixing capacity of the different fractions of the soil. Curr. Sci. 20 : 130-131.

When the clay, silt and sand fractions separately and in various combinations were inoculated and incubated with azotobacter, the greatest N fixation occurred with the clay fraction. Manuring did not affect N fixation by day, but decreased it by sand and manurd soil.

534.
PATI, B.R.; CHANDRA, A.K. and GUPTA, S. (1984) The in vitro effect some pesticides on the nitrogen fixing bacteria isolated from the phyllosphere of some crop plants. Pl. Soil 80 : 215-225.

The in vitro effect of twelve commonly used pesticides including the fungicides ziram and Hinosan, the acaricides Nuvacron, Ekatin 25 EC and Ekalux - G_5 and other insecticides namely

Dimecron, Anthio 25 EC, Baytex, Metasystox and Phosalone on the nitrogen fixing bacteria isolated from the phyllosphere of some crop plants were tested. A very wide variation of the effect of these pesticides, on the different organisms was noted. At a concentration of 700 ug/ml in the medium, most of the pesticides completely inhibited growth of the nitrogen fixing bacterial isolates. However, with some pesticides, when used at a lower concentration, a degree of growth stimulation was recorded.

535.

PATIL, R.B.; GANTOTTI, B.V. and BAGYARAJ, D.J. (1974) Comparative field trials on the performance of different soyabean specific Rhizobium inoculants. Proc. Indian natn. Sci. Acad. 40B : 652-654.

Seven Rhizobium inoculants obtained from different sources were used for treating the seeds of soyabean variety Davis, during "Kharif" 1971. Seed inoculation with Nitragin and I.A.R.I. cultures gave significantly more number of root nodules and also increased grain yield by 26 and 23 per cent respectively. Seed treatment with Bangalore culture recorded 16 per cent increased grain yield compared to control. After soyabean, wheat, variety 'Kalyan' 227 was raise in the same plots in 'Rabi', 1971. With wheat grown after soybean the grain yield increased specially in the plots in which the soyabean seeds inoculated with I.A.R.I. culture was planted. The increase was 29 per cent compared to control.

536.

PATIL, R.B.; PUGASHETTI, B.K. and BAGYARAJ, D.J. (1974) Isolation of efficient strains of Rhizobium for soyabean root nodulation. Proc. Indian natn. Sci. Acad. 30B : 641-643.

The failure of soyabean to establish well in Karnataka was considered to be due to the absence of soyabean-specific indigenous rhizobial strains in the soils. Fourteen strains of rhizobia were isolated from soyabean grown in different parts of Karnataka. A composite culture of the different strains was used for inoculation of soyabean and reisolation of strains was done through three plant passages. Six strains of Rhizobium were thus obtained representing all the three plant passage isolates. All of them appeared to belong to Cow-pea group of rhizobia. Third plant passage isolates proved to be the best, suggesting the possibility of increasing the efficiency of Rhizobium by repeated plant passage.

537.
PATIL, R.B., RAJASHEKHARAPPA, B.J., VISWANATH, D.P. and SHANTARAM, M.V. (1979) Solubilization and immobilization of phosphate by some microorganisms and phosphorus availability to plants.

Bull. Indian Soc. Soil Sci. 12 : 550-556

In both culture experiment, a bacterium (Bacillus megaterium) isolated from cowpea rhizosphere was found more efficiont in solubilizing the rockphosphate than the fungal isolates, namely penicillium, funiculosum and Aspergillus niger. However, Bacillus sp. utilized most of the solubilized phosphorus (P) for its cellular synthesis and released very little of it into the medium. Several organic acids were produced by these microorganisms but their number had no relevance to the phosphate solubilizing capacity indicating that the kind of acid produced was more important. A maximum of 12 to 26 per cent of the P added to soil as labelled tricalcium phosphate was in the available form on 45th day of inoculation. Rock phosphate, together with inoculation, was as good as superphosphate in meeting the requirement of P of cowpea plants. Significantly higher yield of cowpea was obtained with inoculation but an organic amendment, such as farmyard manure, was essential for obtaining maximum solubilization of phosphate by the microorganisms.

538.
PATWARDHAN, P.G.; PRABHU, A.V.; NAGARKAR, M.B. and MAKHIJA, U. (1977) Two terricolous lichens from south western India.

Curr. Sci. 46 : 758-759.

Dermatocarpon hepaticum and Endocarpon pusillumm have been found in association with Riccia spp. on a calcareous soil.

539.
PAUL, N.B. and SUNDARA RAO, W.V.B. (1971) Phosphate-dissolving bacteria in the rhizosphere of some cultivated legumes.

Pl. Soil 35 : 127-132.

Twelve phosphate-solubilizing organisms were isolated from the rhizosphere of 4 cultivated legumes in 4 soil regions of India. The organisms were identified as Bacillus megaterium, B. subtilis, B. brevis, B. pulvifaciens, B. pumilus, and B. polymyxa. The rhizosphers of Trifolium alexandrinum and Phaseolus aureus provided better environments for phosphate dissolving organisms than those of Crotolaria juncea and Sesbania aculeata.

540.
PHILIP, K. and JAUHRI, K.S. (1984) Pressmud : a potential carrier for Rhizobium and Azotobacter. I - Comparative analytical studies of various carrier materials.

Zentb. Microbiol. 139 : 35-41

Pressmud (100%), a waste material of the sugar industry has been found to possess all the desirable physical and chemical properties of an ideal carrier, but it needs amendment with charcoal or soil. Charcoal was found to be the better of the two. Pressmud amended with charcoal (75 : 25) effectively maintained high population of Rhizobium and Azotobacter when it was substituted for the carbon source in broth media.

541.
PILLAI, R.N. and SEN, A. (1966) Salt tolerance of Rhizobium trifolli. Indian J. agric. Sci., 36 : 80-84.

The growth of different rhizobium strains isolated from Trifolium alexandrinum progressively decreased with increasing salinity in the medium; salinitie of 0.5-0.7% completely inhibited growth. Nodulation in plants was not affected by the salinity of the soil, and yields especially of uninoculated plants increased with soil salinity, unlike Vicia faba, the growth of which is adversely affected by salinity.

542.
PILLAI, R.N. and SEN, A. (1969) Effect of salinity on the production of gum by rhizobium species. Indian J. agric. Sci., 39 : 32-35.

A study was made of the effect of salinity on the formation of polysaccharide gums by some strains by rhizobium spp. selected at random. The results showed that gum formation by the organism increased with increasing salinity and that the strains varied in their capacity to form gum in presence of equal amount of salt.

543.
PILLAI, R.N. and SEN, A. (1970) Note on the influence of some trace-element on the growth of Rhizobium from groundnut under stationary culture.

Indian J. agric. Sci. 40 : 1017-1018.

B and Co were not essential for Rhizobium development but their presence enhanced growth. The overall effect of Mn, Zn and Cu on growth was negative. Maximum growth of Rhizobium occurred in the presence of 0.77-3.75 ppm Mo.

544.
PILLAI, R.N. and SEN, A. (1973) Salt tolerance of Rhizobium from Dolichos lablab.

Zentbl. Bakt. ParasitKde. Abt. II, 128 : 538-542.

Six Rhizobium strains grew best with sodium chloride content in the medium at about 0.8 to 1.6%. At higher concentrations growth was reduced.

545.
PILLAI, S.C.; WADHWANI, T.K. and GURBAXANI, M.I. (1947)

Relative efficiency of bacteria and protozoa in the flocculation and oxidation of organic matter suspended in water. Curr. Sci. 16 : 340-341.

Eighty-one bacterial strains isolated from water, soils, compost heaps, raw sewage and sludges and faeces of animals were introduced singly and with protozoa into sterile, aerated suspensions of organic matter prepared from soil, sewage and compost. The bacteria showed poor oxidizing and flocculating activity in comparison with the protozoa and activated sludge. Nitrifying bacteria were present in all the sources examined, but maximum nitrification was only produced in the presence of the protozoa.

546.
PILLAI, S.C.; WADHWANI, T.K. and GURBAXANI, M.I. (1948). Observations on nitrite producing bacteria from different sources and the role of protozoa in nitrification. Curr. Sci. 17 : 122-123.

Nitrite producing bacteria are commonly distributed in nature, but the amounts of nitrite produced by bacteria alone are less than those formed in the presence of protozoa such as *Vorticella* sp. that occur in activated sludge. The addition to composts of small amounts of activated sludge increases nitrification.

547.
POI, S.C., BHADURI, P.N. and KABI, M.C. (1983) Symbiotic effectiveness of gram and lentil in problem soils of West Bengal.

Indian Agric. 27 : 223-228

Application of gypsum to the saline soil and seed pelleting with lime could improve the survival rate, nodulation and yield of both gram and lentil in saline soil. Dolomite treatment of the acid soils improved nodulation, growth and yield of the crop than that of the untreated inoculated control. The yield of gram was poor in acid soil of North Bengal even after the seeds were appropriately inoculated and pelleted with lime. For further improvement, selection of ideal cultivars of pulses for such problem soils have been suggested.

548.
PRABHA, S.; NAIR, S.K.; DADARWAL, K.R. and TAURO, P. (1978) Pseudomonas azotogenesis - an associate symbiont of bajra (Pennisetum americanum).

Pl. Soil. 49 : 657-659.

Bajra was surveyed for associative nitrogen-fixing bacteria in roots. Out of 20 root samples collected from 10 locations, 15 showed a high population of Pseudomonas azotogenesis. The bacterium showed appreciable nitrogenase activity in pure culture.

549.
PRADOSH ROY and MISHRA, A.K. (1981) Factors affecting oxidation of pyrite by Thiobacillus ferrooxidans.

Indian J. exp. Biol. 19 : 728-732

T ferrooxidans, a chemoautotroph, which can utilize reduced iron and sulfur for energy metabolism was used for pyrite oxidation studies. The rate of pyrite oxidation depended on pulp density, initial acidity of the system and acidity increased during pyrite oxidation. A lag period was required during pyrite oxidation by the cells grown on soluble ferrous iron that was eliminated by the application of Tween-80 or by the previous adaptation on pyrite. The oxidation was coupled with the bacterial growth. Most of the cells remained attached with the pyrite particles. In the absence of the essential nutrients, the oxidation was deteriorated. The product of oxidation was soluble iron of which about 90% was in ferric state and sulfide moiety of pyrite was quantitatively oxidized to sulfate.

550.
PRAKASH, D.; NAQVI. Q.A.; ANSARI, A.R.; SAXENA, S.K. and MAHMOOD, K. (1978) Rhizosphere mycoflora of virus-infected and healthy counterparts of Nicotiana glutinosa L.

Fertil. Technol. 15 : 52-53.

The number of fungal species and their abundance in the rhizosphere of N. glutinosa were greater in the case of plants infected with spinach mosaic virus, compared to healthy plants.

551.
PRAKASH, R. and SAKSENA, R.K. (1952) Decomposition of paddy and bajra (Pennisetum typhoideum) straws by fungi commonly found in Allahabad soils. Proc. Indian Acad. Sci. 36 : 119-128.

Decomposition under aerobic laboratory conditions was measured by CO_2 and NH_3 evolution, loss in weight and loss in total C over a 7-day period. Of the 22 species of fungi investigated only 2 were oxidase-positive. The amount of CO_2 evolved was less for some species and greater for others than the loss in weight. The cellulose equivalent of the CO_2 evolved was not proportional to the loss in weight. Decomposition sometimes increased and sometimes decreased the pH of the straw. The rate of decomposition was not related to the pH change of the straw or the oxidase activity of the fungi.

552.
PRAMANIK, B.M. and MISRA, A.N. (1950) Soil bacteria relations at the Shahjhanpur sugarcane research station. Proc. Bienn. Conf. Sug. Cane Res. Workers India. 1 : 97-107.

Manure caused greater increase than ammonium sulphate of equivalent N value in bacterial population of the soil. Fixation of N by Azotobacter showed higher stimulation from sugar factory pressmud than from oil cake or manure, and was reduced by ammonium sulphate.

553.
PRAMANIK, B.N. and MISRA, A.N. (1955) Effect of continuous manuring with artificial fertilizers on Azotobacter and soil fertility. Indian J. agric. Sci. 25 : 1-13.

Continuous application of ammonium sulphate and potassium sulphate depressed N-fixing capacity; a heavy supply of P fertilizers created favourable conditions for azotobacter and other soil microorganisms and thus improved the N-recuperative power of the soils. The higher response in the laboratory to K_2PO_4 than to Na_2HPO_4 showed a capacity for selection by azotobacter strains between P fertilizers of different availability.

554.
PRAMANIK, M. and ISWARAN, V. (1973) Survival of *Rhizobium japonicum* in various carriers.

Zentbl. Bakt. Parasitkde. Abt. II, 128 : 232-239.

Rhizobium japonicum was incubated at 29°C in peat, farmyard manure, charcoal or hay mixed with soil to form 10, 25, 50 and 100 per cent mixtures. The bacterial population was sampled at two week intervals for twelve weeks. The decline in the *Rhizobium* population in carriers with 50% and above of peat or charcoal was not as rapid as in the control soil. Soil amended with FYM was more effective with less than 50% FYM.

555.
PRASAD, B.N. and MEHROTRA, R.K. (1978) Some new additions to Cyanophycean flora of India.

J. Indian bot. Soc. 57 : 98-101

Sixteen members of the Cyanophyceae are described from different types of crop fields of Uttar Pradesh. All these taxa are new records for India.

556.
PRASAD, J. and JHA, K.K. (1968) Protozoa in Bihar soils.

Curr. Sci. 37 : 55-56.

Twenty soil samples contained 157000-8873000 protozoa per g. Population size was affected by altitude, fertilizer applications, and vegetation, and decreased sharply with depth in the top foot of soil.

557.
PRASAD, J. and RAM, H. (1985) Concentration and uptake of Fe by summer mungbean (Vigna radiata) as affected by Zn, Cu and Rhizobium inoculation.

Proc. Natn. Sem. Plant Physiol. (Abst.) p. 113.

Experiments conducted on farmer's field of Varanasi district revealed that Zn and Cu applied @ 2.5 and 5.0 ppm in different combinations with and without Rhizobium inoculation depressed the Fe concentration in plants and grains.

558.
PRASAD, M. (1977) Microbial society and its activity in the soil.

Zentbl. Bakt. ParasitKde Abt. II 132 : 659-665.

In a typical central European garden soil, the buried slide method revealed the presence of fungal mycelia in association with bacterial cocci and rods, individual colonies of Azotobacter-type cells and species of the alga Hantzschia. A nematode attacked by bacteria was revealed by the Cholodny-Strugger method, and this method also revealed fungal mycelia attacked by bacteria. Colonies of rod-shaped Clostridium-type bacteria occurred between soil particles.

559.
PRASAD, N.N. and JAYARAMAN, K.N. (1971) Production of organic acids by certain soil Aspergilli.

Curr. Sci. 20 : 198-199.

Tests with 15 Aspergillus cultures showed that Aspergillus niger produced the most citric, oxalic and gluconic acids, and that kojic acid was produced only by Aspergillus flavus and Aspergillus tamarii.

560.
PRASAD, R. and LAKHDIVE, B.A. (1969) Note on nitrification

of ammonium sulphate and subsequent losses of nitrogen under waterlogged conditions as affected by nitrification retarders "N-serve" and "AM".

Indian J. agric. Sci. 39 : 259-262.

Nitrification of $(NH_4)_2SO_4$, incubated with a sandy clay-loam (pH 8.2) under waterlogged conditions, was effectively retarded on application of N-serve or 2-amino-4 chloro-6-methyl-pyrimidine at 2-5 ppm of soil, thus preventing heavy losses of up to 81% N under water-logging for 96 hours.

561.
PRASAD, T. and SINHA, S. (1973) Survival of Fusarium oxysporum f. lathyri in the decomposing roots of Khesari (Lathyrus sativus).

Indian Phytopath. 26 : 79-83.

Fungi were isolated from decomposing roots and rhizoplane of Lathyrus sativus at monthly intervals for nine months, 19 species representing 13 genera were found but only Fusarium oxysporum f. lathyri was present at all stages of root decomposition.

562.
PURUSHOTHAMAN, D. and BALARAMAN, K. (1973) Effect of soil phenolics on the growth of Rhizobium.

Curr. Sci. 42 : 507-508.

A sample of lignite contained 0.421 µg total phenols/g ovendry soil, one of black soil 0.127 µg, one of peat soil 0.254 µ and one of the red soil 0.246 µg. The lignite contained p-hydroxy benzoic acid, p-aminobenzoic acid, p-coumaric acid, syringic acid and cinnamic acid. The black soil contained only p-hydroxy benzoic acid and p-coumaric acid, and the peat and red soils contained p-hydroxybenzoic acid, p-coumaric acid and syringic acid. Inhibition of growth of Rhizobium in vitro was greater by p-aminobenzoic acid or hydroquinone than by anthranilic acid, cinnamic acid, ferulic acid, gallic acid or p-hydroxybenzoic acid. All these compounds were inhibitory at 0.0005 M concentration.

563.
PURUSHOTHAMAN, D. and GUNASEKHARAN, S. (1979) Does soil inhibit biological nitrogen fixation? Curr. Sci. 48 : 650-652.

The nitrogenase activities of *Azotobacter chroococcum*, *Beijerinckia indica* and *Derxia gummosa* were markedly reduced when incubated with clay loam or coarse sand, in comparison with soil-free controls. The effect of the clay loam was greater than that of the sand. The presence of the adsorbents Dowex-50, Amberlite and activated charcoal also profoundly reduced nitrogenase activity. It is concluded that adsorption of bacterial cells by soil particles is responsible for nitrogenase inhibition.

564.
PURUSHOTHAMAN, D.; BALASUNDARAM, C.S. and GUNASEKHARAN, S. (1981) Nitrogen fixation in the rhizosphere of rice.

Proc. natn. Sci. Acad. 47 B : 242-247.

Azotobacter and *Beijerinckia* were more prevalent in the rhizosphere than in the non-rhizosphere region of the two rice varieties tested (IR 8 and Bhavani). At boot-leaf and flowering stages, the population of these bacteria was very high. In general the numbers of these bacteria were greater in IR 8 than in Bhavani. In IR 8, maximum clostridial population was recorded at flowering whereas in Bhavani it occurred at tillering.

565.
PURUSHOTHAMAN, D.; DHANPAL, N. and RANGASWAMI, G. (1977) Nitrogen fixation under rice field conditions.

Curr. Sci. 46 : 863-864.

N fixation (as measured by acetylene reduction) under two rice varieties was greater in flooded soil than in dry soil, was greater in the light than in the dark, and was greater in the rhizoplane and rhizosphere soil than in non-rhizospher soil.

566.
PURUSHOTHAMAN, D.; MARIMUTHU, T.; VENKATARAMANAN, C.V. and KESAVAN, R. (1974) Role of Actinomycetes in the biosynthesis of indole acetic acid in soil.

Curr. Sci. 43 : 413-414.

Up to 1181/ug IAA/100 ml of culture filtrate was synthesized by isolates of actinomycetes. IAA synthesis in sterile and non-sterile red soil inoculated with actinomycetes and containing dextrose was 12.7 and 28.6/ug/25 ml of soil extract, respectively. When tryptophan was added the corresponding figures were 64.9 and 172.2/ug.

567.
RAGHU, J.S. and CHOUBEY, S.D. (1983) Response of gram to irrigation and fertilization.

Indian J. Agron. 28 : 239-242.

Application of one irrigation 45 days after sowing significantly increased yield of gram as compared with no irrigation and one irrigation 75 or 90 days after sowing. Two irrigations 45 and 75 days after sowing gave the maximum yield. The maximum water use efficiency was due to no irrigation (control) followed by one 45 days after sowing. The seed treatment with Rhizobium culture + 25 kg N + 50 kg P_2O_5/ha significantly enhanced gram yield.

568.
RAI, A.K. and PANDEY, G.P. (1981) Influence of environmental stress on the germination of Anabaena vaginicola akinetes.

Ann. Bot. 48 : 361-370

Germination of akinetes of Anabaena vaginicola v. fertilissima in response to environmental stress was studied. Addition of nitrate to the medium induced early and maximum germination (96%), whereas less than half of the akinetes germinated when either nitrate or phosphate was omitted from the medium. The pH range over which germination occurred was 7.0-9.0. Desiccation and subsequent rehydration of akinetes resulted in a lag period before germination, and a decrease in the germination rate, which were directly proportional to the length of the desiccation period. The temperature optimum for germination and vegetative growth was the same ($25^{o}C$); germination did not occur at $5^{o}C$ or above $35^{o}C$. The limit of heat shock tolerated was $55^{o}C$ for 4 min. In addition to a white light, only the red part of the visible spectrum induced germination. Ultraviolet radiation reduced germination rate. The photoreactivating system in akinetes is non-photosynthetic. LD_{50} photon flux densities were 300 Jm^{-2} for akinetes and 240 Jm^{-2} for vegetative cells.

569.

RAI, A.K. and TALPASAYI, E.R.S. (1978) A Comparative study of nitrate reductase activity of some strains of a blue-green alga Anabaena ambigua Rao. Pp297. In S.P. Sen; Y.P. Abrol and S.K. Sinha (eds.), Nitrogen Assimilation and Crop Productivity. Associated Publishing Company, New Delhi.

In the present investigation, authors have made a comparative study of several strains of a nitrogen-fixing blue-green alga, Anabaena ambigua Rao and used nitrate reduction along with other morphological and physiological features as a basis for characterising the different strains of the same alga.

570.

RAI, B. and SRIVASTAVA, A.K. (1983) Decomposition and competitive colonization of leaf litter by fungi. Soil Biol. Biochem. 15 : 115-117.

The decomposition of mixed leaf litter from a tropical dry-deciduous forest by ten dominant letter-soil inhabiting fungi was studied in vitro. The activity of all the species, measured by CO_2 evolution, was greater for unsterilized than for sterilized litter. The relative activities of the species differed between sterile and unsterile litter.

571.
RAI, B. and UPADHYAY, R.S. (1983) Competitive saprophytic colonization of pigeon-pea substrate by Fusarium udum in relation to environmental factors, chemical treatment and microbial antagonism. Soil Biol. Biochem. 15 : 187-191.

The competitive saprophytic colonization (CSC) of pigeon-pea substrate by Fusarium udum was studied in relation to environmental factors, chemical treatments and microbial antagonism. The optimal temperature for CSC was 22°C, while at 40°C the substrate was vigorously colonized by Coprinus lagopus and Aspergillus nidulans which suppressed colonization by F. udum. High competitive saprophytic ability of F. udum was noted at pH 7 and 9 at soil moisture contents between 5 and 30%.

572.
RAI, B.; SINGH, V.N. and SINGH, D.B. (1980) Trichoderma viride as a mycoparasite of Aspergillus spp. Pl. Soil 57 : 131-135.

Trichoderma viride was parasitic on three species of Aspergillus. The mycoparasitism was characterized by frequent coiling, penetration and hyphal growth of the parasite inside the conidiophores of Aspergillus. The volatile and non-volatile metabolites of T. viride. inhibited radial growth of five Aspergillus spp.

573.
RAI, B.; SRIVASTAVA, A.K. and SINGH, D.B. (1981) Volatile and non-volatile metabolites of actinomycetes and the growth of some litter decomposing fungi. Soil Biol. Biochem. 13 : 75-76.

In laboratory incubations the volatile and non-volatile metabolites of six dominant forest soil and litter actinomycetes (Micromonospora elongata, M. parva, Streptomyces armillatus, S. griseolus, S. griseus and S. rochei) inhibited the growth of the dominant litter decomposing fungi *Aspergillus niger, Penicilluim rubrum* and *Trichoderma harzianum* and stimulated *Curvularia lunata* var. *aeria*; *Acremonium furcatum* was inhibited by the volatile metabolites of all the actinomycetes tested, but stimulated by the non-volatile metabolites of four actinomycete species; *Cladosporium cladosporioides* was stimulated by two and inhibited by four of the actinomycetes. *Streptomyces* spp. appeared to be more active antagonists than *Micromonospora* spp.

574.
RAI, J.N. and TEWARI, J.P. (1961) Addition to our knowledge of Indian soil fungi. *Proc. Indian Acad. Sci.* 54B : 209-217.

Descriptions are given of : *Ascotricha xylina, Tritirachium dependens, Koorchalomella oryzae* (these three first isolated from soil), *Aspergillus terreuse* var. *africanus, Penicillium stepitatum, Gliocladium penicilloides* and two species of *Sclerotium* (all new for Indian soils).

575.
RAI, J.N. and TEWARI, J.P. (1963) A new *Acremoniella* from Indian soils. *Can. J. Bot.* 41 : 331-334.

A. serpentina spec. nov. separated from soil (pH 6.8) is described.

576.
RAI, J.N. and TEWARI, J.P. (1963). Studies in cellular slime moulds from Indian soils. III - On the occurrence of two strains of *Dictiostelium mucoroides* Complex, conforming to the species *Dictiostelium sphaerocephalum* (Oud.) Saccardo and March. *Proc. Indian Acad. Sci.* 58 : 263-266.

Two strains of *D. mucoroides* Complex are described for the first time from Indian soils.

577.
RAI, J.N.; TEWARI, J.P. and AGARWAL, S.C. (1968) A new species of *Aspergillus versicolor* group - A. *Lucknowensis* spec. nov., from Indian alkaline soils.

Can. J. Bot. 46 : 1483-1485.

The species described occurs in "usar" soils of pH 8.

578.
RAI, J.N.; TEWARI, J.P. and MUKERJI, K.G. (1961) A new *Heliocostylum* from Indian soils. Can. J. Bot. 39 : 1281-1285.

A new species of mucoraceous fungus named *Heliocostylum lucknowense,* isolated from soils of pH 6-8, is described.

579.
RAI, J.N.; TEWARI, J.P. and MUKERJI, K.G. (1964) A new *Aspergillus* from Indian soils : *A. striatus* spec. nov. Can. J. Bot. 42 : 1521-1524.

This new member of the *A. nidulans* group was isolated from mangrove mud.

580.
RAI, J.N.; TEWARI, J.P. and MUKERJI, K.G. (1964) *Achaetomium,* a new genus of ascomycetes. Can . J. Bot. 42 : 693-697.

A. globosum n. sp., *A. stumarium,* n. sp. and *A. luteum* n. sp. described were isolated from Indian soils.

581.
RAI, R. (1983) Salt tolerance of rhizobium strains and lentil genotypes and the effect of salinity on aspects of symbotic N-fixation. J. agric. Sci. 100 : 81-86.

Three Rhizobium strains and four lentil genotypes were more salt tolerant under different stress conditions. Saline resistant strains were more antibiotic resistant and showed higher relative rates of oxidation of carbohydrate and tricarboxylic acid intermediates. Under salt stress, significant interactions between strains and genotypes resulted in differential response of N_2 fixation and glutamine synthetase activity.

582.
RAI, R. and PRASAD, V. (1983) Salinity tolerance of *Rhizobium*

mutants : growth and relative efficiency o mb ɔtic nitrogen fixation. Soil Biol. Biochem. 15 : 217-219.

Nitrosoguanidine was used to produce 5 mutant strains of the Rhizobium strain RL 5. The salinity tolerance, streptomycin resistance, growth, nodulation and efficiency of symbiotic N_2-fixation with lentil of these strains were studied. Among the mutants and parent, LM 4 and LM 1 tolerated 200 µg/cc streptomycin and 1.5% NaCl. These two mutants increased the number and dry weight of nodules/plant, the dry matter yield of the crop and N_2-fixation.

583.
RAI, R. and SINGH, S.N. (1979) Response of strains of Rhizobium on nodulation, grain yield, protein and amino acids content of chick pea (Cicer arietinum Linn.). J. agric. Sci. 93 : 47-49.

Nine strains of Rhizobium sp. were studied in vivo for their nodulation capacity, leghaemoglobin content, grain yield, crude protein and 16 amino acids content, in the chick pea variety C 235 grown on a calcareous saline alkali soil. There was no significant correlation between grain yield and number of nodules (r = 0.37) or dry weight of nodules (r = 0.29), but grain yield was significantly correlated with leghaemoglobin content of nodulues (r = 0.95). Of the 16 amino acids analysed in seed samples, aspartic, glutamic, proline and histidine were greatest with strain H 45; glycine, leucine and arginine with strain F 6; norleucine, tyrosine and phyenylanine with strain KG 38; and alanine and valine were treatest with strain KG 41. Strain KG 38 led to significantly higher grain yield than the other strains.

584.
RAI, R.; SINGH, S.N.; SINHA, R.B.; PRASAD, V.; PRASAD, T.N. and SINGH, N. (1984) Effect of pyrites-pressmud complex on symbiotic nitrogen fixation, grain yield and quality of pea grown in calcareous soil.

Indian J. agric. Sci. 54 : 478-481.

Pyrites-pressmud complex significantly increased grain yield, nodulation, leghaemoglobin content, nitrogenase activity of nodules and total carbohydrate of nodules of pea Pisum sativum Linn. var arvense Poir when compared with the control and pyrites or pressmud alone. The crude protein, methionine, cystein, tryptophan, calcium, phosphorus and iron contents of grain also increased significantly. The best combination was 16 q pyrites + 80 q pressmud or 24 q pyrites + 80 q pressmud per hectare.

585.
RAI, S.N. and GAUR, A.C. (1982) Nitrogen fixation by Azospirillum spp. and effect of Azospirillum lipoferum on the yield and N uptake of wheat crop. Pl. Soil 69 : 233-237.

Among 7 isolates of A. brasilense and three of A. lipoferum, those having a greater denitrifying capacity fixed less N in N-free semi-solid malate medium. One strain of A. lipoferum having N-fixing capacity and showing no denitrification was tested as an inoculant to field-grown wheat with different N rates. The treatment receiving 80 kg N/ha yielded 2.97 t/ha as against yield of 4.15 t/ha in a treatment receiving inoculant with fertilizers.

586.
RAJAGOPALAN, N. and SADASIVAN, T.S. (1964) Some aspects of root nodulation in tropical legumes. Curr. Sci. 33 : 197-202.

A discussion of the association between rhizobia and legumes; the effect of soil type and soil reaction on nodulation; the legume-rhizosphere effect; the effect of plant exudates, effective and ineffective rhizobia, effective and ineffective root nodules; haemoglobin content of root nodules; free amino-acids and amides in root nodules; the practical aspects of legume nodulation to tropical agriculture.

587.
RAJAGOPALAN, S. and RAMAKRISHNAN, K. (1963) Phycomycetes in agricultural soils with special reference to Pythiaceae. I. Techniques of isolation.

J. Madras Univ. 33B : 311-341.

Baiting and agar-block techniques were used on rainfed soil, garden land, wet land and semi-wet land. The greatest number of Pythiaceae were isolated in August.

588.
RAJU, G.N.S. and MUKHOPADHYAY, A.K. (1974) Studies on availability of fixed NH_4^+ to nitrifying organisms.

Pl. Soil 41 : 287-291.

In eighteen soils from West Bengal, 5.5-47.9% of fixed NH_4^+ was nitrified during 45 days' incubation at 28^oC; most nitrification occurred between days 15 and 30. Availability of fixed NH_4 was inversely correlated with NH_4 fixing capacity.

589.
RAJU, P.N. (1974) Dolichos enation mosaic virus (DEMV) and symbiotic nitrogen fixation in field bean (Dolichos lablab Linn.) Proc. Indian natn. Sci. Acad. 40B : 629-635.

Dolichos Enation Mosaic Virus (DEMV) infection decreased the levels of total carbohydrates and reducing sugars in roots and nodules of D. Lablab after 26 days of growth. However, the nitrogen content in these nodulated virus infected plants was more than that in the healthy plants. Again, nodules of virus infected plants contained higher amounts of total nitrogen after 24 days. There was a higher rate of transfer of fixed nitrogen from nodules to the tops and roots as virus infection of the host soluble proteins, total soluble nitrogen constituents and the total insoluble nitrogen increased in nodules following virus infection of the legumes. Nitrogen fixation in infected plants increased during 24 to 31 days to levels higher than that in healthy plants and this high level was maintained throughout the growth period. An increased nitrogen fixing activity to symbiosis rhizobia in nodules of D. lablab following DEMV infection of indicated.

590.
RAM, G. and SANGRIA, C.L. (1979) Effect of seed inoculation

on peas (Pisum sativum) with Rhizobium and Azotobacter in alkali soil. J. Indian Soc. Soil Sci. 27 : 489-491.

Based on the yield response to cultures in two experiments, treatment Rhizobium F_{10}+Azotobacter B_{5}, seemed promising for growing pea in alkali soils with pH around 9.

591.
RAMA RAO, P. (1964) Curvularia tuberculata from dune soils. Curr. Sci. 33 : 121.

The fungus described was isolated from coastal dunes by the soil plate method.

592.
RAMACHANDRA REDDY, T.K. (1959) Foliar spray of urea and rhizosphere microflora of rice (Oryza sativa L.). Phytopath Z. 36 : 286-289.

In pot experiments with rice soil, two and five sprayings of the crop with 30 ml of a 0.1 M solution of urea caused a preferential stimulation of Penicillium sp. with a resulting reduction in the root zone. Since the soil was protected against contact with spray, it seems likely that urea absorbed by the leaves changed the nature and amount of root exudates, thus affecting the microbial balance in the rhizosphere.

593.
RAMACHANDRA REDDY, T.K. (1959) Rhizosphere microflora of pteridophytes. Curr. Sci. 28 : 113-114.

Quantities of fungi and bacteria in the rhizosphere of Lastrea, Adiantum, Pteris and Equisetum decreased in that order. There were significant increase in Penicillium and Aspergillus species in the region of the roots; Trichoderma sp. were frequently isolated but pathogenic forms of Fusarium were infrequent.

594.
RAMACHANDRA REDDY, T.K. (1961) A new species of Didynosphaera from the rhizosphere. Mycologia 53 : 471-473.

The fungus was isolated from rhizosphere soil of rice at Madras.

595.
RAMACHANDRA REDDY, T.K. (1968) Plant treatment in relation to the rhizosphere effect. I. Pretreatment of roots of rice seedlings to certain chemicals and antibiotics on the rhizosphere micro flora.

Pl. Soil 28 : 347-356.

Rice seedlings were soaked in 1 and 5 ppm of patulin, griseofulvin, actidione and gibberellin, 5, 10 and 50 ppm of agrimycin-100, 0.1 M urea and distilled water; they were then grown in rice-field soil for a week. Quantitative analyses of bacteria, fungi and actinomycetes of the rhizosphere were made. 1 ppm of griseofulvin, gibberellin and patulin reduced the numbers of bacteria and fungi, but only slightly affected actinomycetes. Griseofulvin had a greater effect on fungi than other chemicals did. Bacteria and fungi were even more affected by 5 ppm, but actinomycetes increased in numbers. Actidione and griseofulvin retarded root growth. Agrimycin-100 at 5 and 10 ppm reduced the rhizosphere effect, but at 50 ppm, fungi, bacteria and actinomycetes greatly increased in numbers. 0.1 M urea had little effect on bacteria or fungi, but numbers of actinomycetes increased.

596.
RAMACHANDRA REDDY, T.K. (1968) Plant treatment in relation to the rhizosphere effect. II. Foliar application of certain chemicals and antibiotics in relation to the rhizosphere microflora of rice (Oryza sativa L.).

Pl. Soil 29 : 102-113.

Foliar sprays of different nutrients and antibiotics affected the rhizosphere microflora. The type and magnitude of the effect depended upon the nature and concentration of the substances and varied with different seasons.

597.
RAMACHANDRA REDDY, T.K. (1968) Plant treatment in relation to the rhizosphere effect. III. Foliar application of certain trace elements and metallic chelates in relation to rhizosphere microflora of rice (<u>Oryza sativa</u> L.).

<u>Pl. Soil</u> 29 : 114-118.

Foliar application of trace elements (Zn, B, Mn and Mo) and trace element chelates (Zn, Fe, Cu and Mo) to three-week old rice seedlings had pronounced effects on the distribution of rhizosphere microflora as well as the percentage occurrence of certain soil fungi. In the absence of direct evidence on the influence of sprays on the physiology and root-exudate pattern, it is suspected that the rhizosphere effect is due to possible changes in root exudates in consequence of foliar sprays.

598.
RAMAKRISHNA, C.; GOWDA, T.K.S. and SETHUNATHAN, N. (1979) Effect of benomyl and its hydrolysis products, MBC and AB, on nitrification in a flooded soil.

<u>Bull. Environ. Contamin. Toxicol.</u> 21 : 328-333.

Benomyl at 10 and 100 ppm did not greatly affect nitrification of ammonium sulphate in 10-g samples of an alluvial soil maintained under flooded conditions. However, 1000 ppm prevented nitrate formation. MBC at 1000 ppm caused slight inhibition while AB (2-aminobenzimidazole) caused some inhibition at 100 and 1000 ppm. Benomyl inhibited nitrite formation by pure cultures of <u>Nitrosomonas</u> sp. and nitrate formation by <u>Nitrobacter agilis</u> at 10 ppm; AB was inhibitory at 100 ppm, while MBC was not inhibitory.

599.
RAMAKRISHNA, C.; RAO, V.R. and SETHUNATHAN, N. (1978) Nitrification in simulated oxidized surface of a flooded soil amended with carbofuran.

Soil Biol. Biochem. 10 : 555-556.

Tabulated data describing the oxidation of carbofuran in the stimulated oxidized surface of a flooded soil and the effect of carbofuran on the oxidation of NH_4 by autotrophic bacteria in a mineral salts medium indicate that carbofuran at rates close to those used in the field is not harmful to nitrification, but that a concentration as high as 0.5% could lead to an accumulation of NO_2- as a result of a combination of the selective inhibition of NO_2- oxidation with a slight stimulation of NH_4+ oxidation.

600.
RAMAN RAI; SRIVASTAVA, B.S.; SINGH, B.P.; SINGH, S.N. and PRASAD, V. (1982) Studies on symbiotic N_2-fixation by mutant strains of lentil (Lens esculenta) Rhizobium in acid soils.

Trans. 12th int. Congr. Soil Sci., New Delhi 6 : 53.

A strain of Rhizobium isolated from the root nodule of lentil was mutagenised and screened on a medium of acidic pH (4.5). Five mutant clones were identified which showed their growth capability in medium of pH range 4.5 to 7.5. N_2- fixation by nodules formed by the two strains was studied in acidic and in neutral soils and their relative efficiencies observed. Using the number of nodules, their dry weight, total plant nitrogen, relative efficiency of N_2- fixation and active iron content of nodules, mutant strain S_1 was found to be equally effective in all the soils. However, strain M_1 was found more responsive and effective in acid soils. Not only the soil pH but associated factors of acidity such as Al, Mn and Fe were also found to be responsible for symbiotic N_2- fixation in acid soils.

601.
RAMASWAMI, P.P. and RAJ, D. (1973) Studies on the evolution of carbon dioxide in soils of Tamil Nadu.

Madras agric. J. 60 : 885-889.

In pot experiments CO_2 evolution was higher in soils treated with green manure than in soils treated with N, P or K. Highest amounts were evolved from a lateritic soil, presumably due to its appreciable content of organic matter. Biological activity was maximum at the tillering stage of the rice crop.

602.
RAMASWAMI, P. RAJ, D. (1979) Influence of phosphorus fertilization on P transformation, microbiological activities, nutrient uptake and yield of rice.

Bull. Indian Soc. Soil Sci. 12 : 270-273.

A pot culture experiment with five soil groups was conducted with rice Co 32 strain as a test crop. Phosphorus application increased the phosphorus availability in the soils and was beneficial for enhancing bacterial, actinomycetal and fungal populations. In general, it was observed that conditions of submergence favoured the availability of phosphorus..

Biological activities taking place in the soil viz., CO_2 evolution, nitrogen fixing, ammonifying and nitrifying capacity were not significantly influenced due to phosphorus application. Grain and straw yields of rice were significantly increased due to phosphorus fertilization but not the root growth, Phosphorus and potassium uptake by grain and potassium uptake by straw were enhanced by phosphorus application. The trend with respect to yield of grain, straw and root was similar within the groups of soils studied and was of the decreasing order in lateritic, black and red soils. Available phosphorus content of the soil correlated closely with grain yield.

603.
RAMASWAMI, P.P.; MATHAN, K.K.; and NAIR, K.S. (1977) Azotobacter population in red and black soils of Tamil Nadu.

Mysore J. agric. Sci. 11 : 364-366.

The numbers of Azotobacter were greater in black soils than in red soils. The numbers decreased with depth, especially in the case of black soils. Seasonal variations were not significant.

604.
RAMASWAMI, P.P.; MATHAN, K.K. and SUYAMBULINGAM, C. (1977) Denitrifying bacteria in red and black soils.

Madras agric. J. 64 : 218-221.

Investigations on the seasonal variations in the denitrifying bacterial population and distribution pattern in the various horizons of red and black soils over a twelve month period showed population differences between these soils. The cyclic variation of the denitrifiers was a function of the rainfall distribution. The population was most concentrated in the top 30 cm of soil and decreased with depth to 90 cm. Thereafter, no variation with depth was apparent.

605.
RANGA RAO, V. and SUBBA RAO, N.S. (1974) Factors responsible for non-nodulating nature of some legumes. Proc. Indian natn. Sci. Acad. 40B: 613-617.

The roots and root extracts of the three non-nodulating legumes viz. *Cassia fistula, C. accidentalis* and *Leucaena leucocephata* inhibited the growth of six strains of *Rhizobium, in vitro.* The presence of antagonistic micro-organisms in the root zone and the inhibitors in the roots are suggested as one among the several factors resulting in lack of nodulation in these legumes.

606.
RANGANATHAN, S. and NORRIS, R.V. (1927) Nitrogen fixation by *Azotobacter chroococcum.* J. Indian Inst. Sci. 10A : 79-96.

Cultures of *Azotobacter* preserve their vitality and vigour much better in soil extract-agar-medium containing 1 per cent mannite than in mineral salt-mannite-agar medium. Nitrogen fixation increases in proportion to the amount of sugar fermented in the medium; more than 50 per cent of the N fixed is assimilated within the first few days of the fermentation. There is no regular increase in N-fixation with increasing concentrations of sugar in the medium. The first product of N-fixation is ammonia which is gradually worked up into more complex forms of N through mono-and diamino N compounds. Phosphates accelerate the process of N-fixation but do not economise sugar-consumption.

607.
RANGANATHAN, S. and NORRIS, R.V. (1927) Studies in intensive bacterial oxidation. IV-Factors influencing the rate of oxidation. J. Indian Inst. Sci. 10A : 107-116.

The speed of nitrification decreased with increased concentration of nitrate. With dilute solutions of ammonium salts and with $CaCo_3$ in quantity necessary to neutralize the resulting acid, but employing no nutrient salts, a concentration of 5.56% of calcium nitrate was obtained. Only about 96% of the N added was recovered as nitric nitrogen. The nitrifying organisms thrive in media of pH ranging from 5.8 to 8.0, but function best when the pH lies between 7.0 and 8.0.

608.
RANGASWAMI, G. (1968) Influence of plant growth on soil microflora.

Indian J. agric. Chem. 1 : 25-31.

A discussion of the effects of germinating seeds the rhizosphere, and crop rotations on the microflora of soil.

609.
RANGASWAMI, G. (1980) Influence of plant growth on soil microflora. In Soil Science & Agricultural Chemistry, Part I. Indian Society of Agricultural Chemists, Allahabad. p 144-154.

A review article which stresses that plant species exert different qualitative and quantitative effects on the microflora not only through their varying types of roots but also through their root exudations which changes with age. The studies made so far have indicated that the microflora could be altered through (a) pretreating the seed with chemicals and certain micro-organisms, (b) foliar sprays with some systemic chemicals and (c) through amendments with organic and inorganic manures.

610.
RANGASWAMI, G. (1982) Research and development for biological nitrogen fixation in India. In Biological Nitrogen Fixation Technology for Tropical Agriculture (edited by Graham, P.H. & Harris, S.C.). Centro Internacional de Agricultura Tropical (CIAT) p 683-686.

Interest in BNF research and development in India has increased in recent years because of the gap in nitrogen fertilizer demand and supply. Currently some Rs 2 million is spent annually on BNF research, with Rs. 12 million for development. Emphasis is toward a better understanding of the legume/*Rhizobium* symbiosis, but greater attention has been paid recently to the use of Azolla, N_2-fixing blue-green algae and associative symbiosis. It is hoped that saving in N fertilizer use of as much as Rs. 150 million can be achieved.

611.
RANGASWAMI, G. and NARAYANASWAMI, R. (1965) Studies on the microbial population of irrigation water in rice fields.

Int. Rice Commn. Newsl. 14 : 35-39.

Total populations of bacteria, fungi and actinomycetes present in channel water leading to rice fields and in stagnant water in the rice nursery and main field were estimated periodically. There were large variations in population due to seasonal factors. Rainfall, cultural operations such as plowing and puddling, application of fertilizer, farmyard manure and green leaf manure seem to alter the microbial population greatly. There were changes in population caused by rice growth. There was a direct correlation between solid matter content of the water samples and bacterial populations.

612.
RANGASWAMI, G. and OBLISAMI, G. (1962) Studies on some

legume root nodule bacteria.

J. Indian Soc. Soil Sci. 10 : 175-186.

Fifteen leguminous plants were examined for root nodule bacteria and 26 isolations of rhizobia were made. The infectivity and nodulating capacity of 22 of them on the respective hosts were established. A comparison was made of the number and types of nodules formed on different hosts and the isolates were grouped under various known cross infection groups of Rhizobium.

Detailed cross inoculation studies made with the rhizobia from Sesbani speciosa, Crotalaria retusa and Clitoria ternatea showed that all the three isolates could readily nodulate several known hosts of cowpea group and vice versa. They could also nodulate soyabean plants and vice versa, thus establishing close kinship between the two groups. The variations among the isolates of the cowpea group in the cross infectivity was brought in that the isolate from Crotalaria retusa could also nodulate Frenchbean plants of the bean group and vice versa. The isolates from the three hosts could effectively nodulate the respective hosts and supply the nitrogen requirements of the plants as indicated by the increases in the nitrogen content of the various plant parts due to nodulation.

Even though there were variations and overlappings in the cross infectivity among some of the isolates of a group, all the isolates invariably confirmed to the respective standards of the species in their cultural, physiological and biochemical properties as detailed in the Bergey's Manual.

613.
RANGASWAMI, G. and SADASIVAM, K.V. (1964) Studies on the occurrence of Azotobacter in some soil types.

J. Indian Soc. Soil Sci. 12 : 43-49

Soil samples from places around Chidambaram, Madras State were examined for their total microbial and Azotobacter populations and for some chemical and physical properties. Azotobacter was present invariably in all the soils and in the three layers; the population, however, decreased with depth. There was no apparent correlation between the total microbial and Azotobacter populations in a given soil. A. chroococcum Beij. and A. indicus Beij. and A. lacticogenus Kauf. and Tours. In acid soils, A. indicus was predominant, whereas A. chroococcum was so in alkaline and neutral soils. The most important factor affecting the Azotobacter population was the hydrogen ion concentration of the soil. The presence of organic matter, CaO and P_2O_5 also seemed to influence the number but not nitrogen and K_2O contents. Isolates of A. chroococcum fixed relatively more nitrogen in the liquid medium than those of A. indicus and one of the isolates of the former fixed almost double the quantity, indicating thereby the prevalence of strain variations in the species.

614.
RANGASWAMI, G. and SUBBARAJA, K.T. (1962) Azotobacter in paddy soil. Proc. Indian Acad. Sci. 56B : 174-183.

In order of population strength the species present were A. chroococcum, A. indicus, A. lacticogenus, A. agilis, A. beijerincku and A. vinelandu. Numbers were greater and N fixation was more effective in dry soils. A. chroococcum was most effective, closely followed by A. indicus and A. lacticogenus. A. chroococcum, A. agilis, A. beijerincku and A. vinelandu fixed maximum N at pH 6-7 whereas A. indicus and A. lacticogenus were relatively more active at pH 3-8.

615.
RANGASWAMI, G. and VASANTHARAJAN, V.N. (1962) Studies on the rhizosphere microflora of citrus trees. Can. J. Microbiol. 8 : 473-489.

Bacteria were about 40-90 times, actinomycetes 2-6 times and fungi 3-8 times more abundant in the rhizosphere than in soil under Citrus grandis, C. sinensis and C. reticulata. Gram negative non-sporing rods, NO_3 reducers, gelatine liquefiers starch hydrolyzers and methylene-blue reducers were more numerous in the rhizosphere than in the soil, while gram positive rods, cocci and sporeformers were more abundant in the soil. Antagonistic forms of actinomycetes were more abundant in the rhizosphere than in the soil.

616.
RANGASWAMI, G. and VENKATESAN, R. (1964) The rhizosphere microflora of rice plants as influenced by soil depth and root maturity. Curr. Sci. 33 : 181-183.

The rhizosphere effect tended to increase with depth; in young roots this showed particularly with bacteria and was also seen with actinomycetes but did not apply to fungi. In old roots the increase in bacterial population with depth was definite but less marked than with young roots while the increase in actinomycete and fungal populations were more marked. It appeared that the composition of the microflora was not only altered by selective activation of individual types, but that certain microflora were transported on the root surfaces to deeper levels in the soil.

617.
RANJAN, S. and BUDHRAJA, B. (1959) A study of nitrogen fixation in the detached nodules of sunhemp.

Proc. nat. inst. Sci. India 25 : 39-43.

N uptake by detached nodules starts at a high level and gradually falls to zero. In nodules on small pieces of root, uptake starts at and maintains a very high level, only in attached nodules is there a marked increase in total and amide N.

618.
RAO, A.C.S. (1973) Factors affecting the root distribution pattern of wheat under dry farming conditions.

J. Nucl. Agric. Biol. 2 : 4-5.

Soil bulk density was found to influence the root distribution of wheat under dry farming conditions down to 24 cm depth.

Available moisture at seeding also influenced root distribution.

619.
RAO, A.L.S. (1938) Nitrogen fixation in relation to wavelength.

Curr. Sci. 7 : 59-60.

Garden soil containing sucrose was exposed to light from a 1000 watt bulb for 300-400 hours, using filters. Maximum non-symbiotic fixation of N occurred with yellow light.

620.
RAO, A.S. (1965) Fungal flora of two soils at Tirupati.

Curr. Sci. 34 : 489-490.

Fungi isolated both from a red sandy soil (1.84% organic matter) and black clay soil (7.96% organic matter) included *Cunninghamella echinulata, Aspergillus flavus, Penicillium citrinum,* etc., and sterile forms (dark-coloured forms in red soil and hyaline forms in black soil). Fungi isolated from red soil alone included *Cephalosporium acremonium, Aspergillus fumigatus, Mucor hiemalis,* etc., while those from black soil alone included *Mucor silvaticus, Pestalotia* sp. *Aspergillus candidus,* etc. The great difference in fungal populations of the two soils supports the local belief that the black soil is the more fertile one.

621.
RAO, A.V. and ISWARAN, V. (1973) Effect of phenolic compounds on the growth of *Rhizobia.*

Zentbl. Bakt. ParasitKde. Abt. II. 128 : 753-757.

Catechol, cinnamic acid, vanillic acid and p-coumaric acid inhibited the growth of several rhizobia species. Sensitivity to the different compounds varied with the different species. The inhibitory action was not directly proportional to the concentration of the compound tested.

622.
RAO, A.V. and SHARMA, R.L. (1978) Influence of chemicals on the microflora and enzyme activities in the cauliflower rhizosphere.

Acta Bot. Indica. 6 : 71-74.

Populations of fungi, bacteria, actinomycetes, and free living N-fixing bacteria and the activities of hydrolysing enzymes were estimated in rhizosphere soil samples from plants 24 hours after spraying with benlate, dithane, captan, terramycin or streptomycin. The parallel changes in the enzyme activities and fungal populations brought about by the sprays indicate an interrelationship.

623.
RAO, A.V. and VENKATESWARLU, B. (1983) Microbial ecology of the soils of Indian desert.

Agric. Ecosyst. Environ. 10 : 361-369.

Although different groups of microorganisms eg. fungi, actinomycetes, bacteria, Azotobacter and nitrifying bacteria were present, their numbers were low and varied greatly between sites. Actinomycetes formed a high proportion of the total microflora. In the majority of the soils, 80% of the fungal population belonged to the genera Aspergillus and Penicillium, while Gram-positive spore formers were the dominant group among bacteria. No significant decline in the population of microorganisms was observed during summer, in spite of high surface soil temperatures. Stabilization of shifting sand dunes by the introduction of vegetation has markedly increased the soil microflora. In general the low organic matter content and poor moisture availability of desert soils were the major factors limiting optimum microbial activity.

624.
RAO, A.V. and VENKATESWARLU, B. (1983) Nitrogen fixation in pearlmillet by Azospirillum brasilense. Proc. Natl. Symp. Adv. Front. Pl. Sci. 9 : 103-104.

All the cultures showed identical morphological and physiological characters despite the diversity of their origin. There was no correlation between nitrogenase activity of roots and the efficiency of the strain thus indicating the interaction between the host plant and the bacterial strain in working of the nitrogenase system.

625.
RAO, D.L.N.; VARMA, A.K. and SUBBA RAO, N.S. (1978) Dinitrogen fixation by Azotobacter chroococcum and its isocitrate dehydrogenase activity. pp.215-220. In S.P. Sen, V.P. Abrol and S.K. Sinha (eds.), Nitrogen Assimilation and Crop Productivity, Associated Publishing Company, New Delhi.

Azotobacter chroococcum strains were isolated from the rhizosphere of sugarcane and their capacity to fix nitrogen and produce growth promoting substances like indoles and gibberellins were studied. An attempt was also made to study the isocitrate dehydrogenase activity and relate it with their N_2 fixation ability.

626.
RAO, J.L.N. and RAO, V.R. (1983) Nitrogenase activity in the rice rhizosphere soil as affected by Azospirillum inoculation and fertilizer nitrogen under upland conditions. Curr. Sci. 52 : 686-688.

In a field trial during wet season, Azospirillum inoculation slightly increased the rhizosphere nitrogenase activity for up to 88 days after sowing. After urea treatment of the soil the effect was greater. The populations of anaerobic nitrogen fixers and Azotobacter were found to increase after inoculation + urea treatment but no change in the number of Azospirillum was observed.

627.
RAO, J.L.N.; PASALU, I.C. and RAJARAMAMOHAN RAO, V. (1983) Nitrogen fixation (C_2H_2 reduction) in the rice rhizosphere soil as influenced by pesticides and methods of their application. J. agric. Sci. UK 100 : 637-642.

A range of insecticides was applied to rice in a field experiment by incorporation into the soil, by application to the standing water or as a seedling root dip. Effects on nitrogenase activity and nitrogen-fixing microorganism population in the rhizosphere varied according to the insecticed used, the mode of application and the age of the plants. Both stimulation and inhibition were observed.

628.
RAO, J.L.N.; PRASAD, J.S. and RAJARAMAMOHAN RAO, V. (1984) Nitrogen fixation (C_2H_4 reduction) in the rice rhizosphere as influenced by pesticides and fertilizer nitrogen.

Curr. Sci. 53 : 1155-1157

Application of carbofuran effected significant stimulation in many samplings. Prophos, essentially a non-systemic nematicide, exerted considerable stimulation of rhizosphere soil nitrogenase. Another nematicide methan sodium also stimulated soil nitrogenase but to a lesser extent compared to prophos. Fertilizer nitrogen when applied in levels beyond 40 kg N/ha inhibited nitrogen fixation.

629.
RAO, J.L.N.; REDDY, B.B. and RAJARAMAMOHAN RAO, V. (1983) Rhizosphere soil nitrogenase (C_2H_2 reduction) as influenced by the nitrogen management in intermediate deep water rice. J. agric. Sci., U.K. 101 : 547-551.

There was a significant reduction in nitrogenase activity when urea was applied to shallow water as briquettes, but when it was applied behind the plough or between the rows, nitrogenase activity was stimulated. An increased water level of about 50 cm for prolonged periods considerably reduced nitrogenase activity.

630.
RAO, K.V.N. (1963) Determination of optimum pH range for the growth of Oocystis marssonu Lemm. in three media differing in nitrogen source. Indian J. Pl. Physiol. 6 : 135-141.

Optimum pH range for algal growth was 6-8 in all culture media. Growth was generally depressed at pH < 5.5, chlorophyll synthesis was unaffected by low pH in presence of NH_4 or urea as N source but was suppressed at pH < 5 in presence of NO_3.

631.
RAO, K.V.N. (1963) The effect of molybdenum on the growth of Oocystis marssonii Lemm. Indian J. Pl. Physion 6 : 142-149.

The presence of 10^{-7} g/litre Mo in a growth medium containing NO_3 as sole source of N was indispensible for algal growth; Mo deficiency inhibited chlorophyll synthesis but different concentrations of Mo had no marked effect on NO_3 absorption.

632.
RAO, P.S.N. and TALPASAYI, E.R.S. (1978) Effect of salt concentration on nitrate reductase of some brine algae. pp.298-299. In S.P. Sen, Y.P. Abrol and A.K. Sinha (eds.), Nitrogen Assimilation and Crop Productivity, Associated Publishing Company, New Delhi.

From the salt brines we have isolated two N_2-fixing blue-green algae, Anabaenopsis arnoldii and Anabaena fertilissima and two green algae, Schroderia setigera (Lemm) Smith and Dunaliella salina. Among the four brine algae A. arnoldii dominates others in the reserviors and condensors for a few months after rains giving blue-green colour to the brine. In the present investigation we report nitrate reducase activity of the above algae under salt stress.

Enzyme activity increased up to 0.34 M and 1.02 M salt concentration in A. fertilissima and A. arnoldii, respectively, and thereafter showed a decreasing trend. However, presence of salt did not stimulate nitrate reducase in either of the green algae, instead, it brought about a gradual decrease with the increasing salt concentration. This finding indicates that the enzyme in the above blue-green algae is perhaps salt stimulated one.

633.
RAO, V.M.; KANT, S.; RATNAM, B.V. and NARAYANA, H.S. (1978)

Effect of indole acetic acid (IAA) on the rhizosphere micropopulation of Trigonela foenumgraecum Linn.

J. Indian Bot. Soc. 57 : 83-86.

Seeds of Trigonella foenum-graecum were soaked for 9 hours in different concentrations of IAA. A pot experiment was carried out for 45 days and the rhizosphere soil was plated out by dilution plate technique for the study of both bacterial and fungal micropopulation. Increase in the rhizosphere micropopulation was observed in the plants treated with hormone as compared to the control. Amongst the treated plants increase in micropopulation was observed with the increase in the hormone concentration of IAA up to 100 ppm, beyond which there was fall in number. Qualitative differences in the rhizosphere micropopulation was also observed in the hormone treated plants.

634.
RAO, V.R. (1976) Nitrogenase activity in Rhizobium associated with leguminous and non-leguminous tissue cultures.

Pl. Sci. Letters 6 : 77-83.

Nitrogenase activity was observed in root, stem-and leaf-tissue cultures of three genetic lines of Trifolium pratense infected with Rhizobium trifolii Rothamsted Strain 5, Stylosanthes gracilis with Rhizobium sp. Strain CH1552 and root-tissue cultures of carrot with Strains CH1552 and NGR99. Rhizobium (CB1552) previously associated with Stylosanthes tissues continued to show nitrogenase activity for 5 days after separation from the callus. Infected Trifolium callus produced roots in 8 weeks on a medium free from NO_3^-, 2,4-dichlorophenoxyacetic acid (2, 4-D) and kinetin, whereas the uninfected callus remained undifferentiated.

635.
RAO, V.R. (1978) Effect of carbon sources on asymbiotic nitrogen fixation in a paddy soil.

Soil Biol. Biochem. 10 : 319-321.

The effect of various carbon sources on asymbiotic N_2 fixation in a paddy soil was studied by acetylene reduction and ^{15}N tracer techniques. N_2 fixation increased with increasing concentrations of cellulose and rice straw, but the former was more effective. N_2 fixation was considerably enhanced in soils amended with glucose, sucrose, succinate, acetate, butyrate, pyruvate or n-propanol, while n-butanol, ethanol, lognin and paraffin were ineffective. Succinate at low moisture content and butyrate under flooded conditions stimulated N_2 fixation during a 15-day incubation.

636.
RAO, V.R. and MODI, V.V. (1968) Nitrogen enrichment in sludge by bacterization.

Curr. Sci. 37 : 407.

Incubation of Azotobacter chroococcum for six days at $30^{o}C$ on a rotary shaker containing a 1 g sample of poultry-farm sludge (digested or undigested) in 50 ml of sucrose-salt medium increased the N content of the sludge.

637.
RAO, V.R. and RAO, N.S.S. (1974) Factors responsible for non-nodulating nature of some legumes.

Proc. Indian Nat. Sci. Acad. 40B : 613-617.

The roots and root extracts of Cassia fistula, C. accidentalis and Leucaena leucocephala inhibited the growth in vitro of six strains of Rhizobium.

638.
RAO, V.R.; CHARYULU, P.B.B.N.; NAYAK, D.N. and RAMAKRISHNA, C. (1978) Nitrogen fixation by free-living organisms in tropical rice soils.

Basic Life Sci. 10 : 354.

$15N_2$ and acetylene reduction techniques were used in studies for determining the overall contribution by free-living bacteria in the N economy of tropical rice soil amended with rice straw and mineral fertilizers. Application of rice straw significantly enhanced the N-fixing potential in a paddy soil; active N_2 fixation was noticed up to the flowering stage of rice with a max, activity at the tillering stage. The $15N_2$ technique produced lower values for N fixation than the acetylene reduction technique. Spirillum spp. were identified in all the soil types studied.

639.
RAO, V.R.; JAYAKAR, M.; SHARMA, K.R. and MUKERJI, K.G. (1972) Effect of foliar spray of morphactin on fungi in the root zone of Capsicum annuum.

Pl. Soil 37 : 179-182.

A foliar spray of morphactin at concentrations up to 1000 ppm decreased the root mycoflora population of plants growing in pots, probably due to retardation of lateral root growth.

640.
RAO, W.V.B.S. (1946) Soil microbiology and bearing on concepts of fertility and mineral deficiency.

J. Univ. Bombay 15 : 45-47.

The higher temperature in July promotes bacterial activity, with a resulting depletion of the organic matter of the soil. The element lacking in a soil is determined by incubation in a synthetic medium at 34° and determination of the increased CO_2 production after various additions.

641.
RAO, W.V.B.S. (1971) Field experiments on nitrogen fixation by nodulated legumes.

Pl. Soil : 287-291.

P increased N uptake by lucerne on a saline soil. N fertilizer or inoculation with Rhizobium meliloti did not increase yields.

In soils with no indigenous R. japonicum, inoculation increased soyabean yields and the additional N uptake was 40-120 kg/ha. Gram and groundnut also responded to Rhizobium inoculation in field trials.

642.
RAO, W.V.B.S.; KRISHNAMURTI, P.V. and RAO, G.G. (1938) Mechanism of the biological oxidation of ammonia. I - Formation of intermediate products. J. Indian Chem. Soc. 15 : 599-603.

During the oxidation of NH_3 to nitrite by bacteria from various Indian soils neither hydroxylamine nor hyponitrite could be detected. The Corbet test does not appear to be specific for hyponitrite.

643.
RATHI, K.S. and VAISHYA, R.D. (1983) Response of lucerne to phosphorus and rhizobium inoculation.

Indian J. Agron. 28 : 315-316.

Application of phosphorus increased the fodder yield of lucerne in all the 14 cuttings. Seed inoculation with rhizobium melilotii increased the fodder yield of lucerne significantly in all the cuttings and the total yield.

644.
RAY, R.C. (1983) Toxicity of the pesticides hexachlorocyclohexane and benomyl to nitrifying bacteria in flooded autoclaved soil and in culture media. Environ. Pollut. 32 : 147-155.

The influence of HCN and benomyl on nitrification in a flooded autoclaved soil was compared with that in culture media inoculated with Nitrosomonas and Nitrobacter spp. The rate of nitrification was slow in autoclaved soil and faster in culture media. Both the pesticides inhibited nitrification at concentrations of 5 ppm and above. Commercial formulations of HCN and benomyl were more toxic to nitrifying bacteria than the technical grade, probably owing to interactions of the carriers and pesticides.

645.
RAY, R.C. (1984) Effect of some pesticide combinations on sulphur oxidation in the simulated oxidised layer of a flooded soil.

Environ. Pollut. 34 (1) : 15-22.

When applied singly the pesticides did not affect S-oxidation but some combinations, including benomyl + carbofuran, caused a synergistic decrease in sulphate formation from elemental S, reflected in higher soil pH samples and a proportionate decrease in the population of S-oxidising autotrophic bacteria.

646.
RAY, R.C.; RAMAKRISHNA, C. and SETHUNATHAN, N. (1980) Nitrification inhibition in a flooded soil by hexachlorocyclohexane and carbofuran. Pl. Soil 56 : 156-168.

Oxidation of ammonium to nitrate in a flooded soil was inhibited significantly by 10 ppm a.i. of hexachlorocyclohexane (HCH) and almost completely by 100 ppm a.i., concomitant with a proportional decrease in the populations of ammonium-and nitrite-oxidising autotrophic bacteria. A strong synergistic effect on the inhibition of nitrification was observed when HCH and carbofuran were applied in combination.

647.
REDDY, G.R. and DATTA, N.P. (1965) Use of dicyandiamide in nitrogen fertilizers.

J. Indian Soc. Soil Sci. 13 : 135-139.

Dicyandiamide application inhibited the nitrification of ammonium nitrogen in fertilizers added to soil. Application of compost in combination partially counteracted this inhibitory effect. Inhibitory effect of the compound on different fertilizers was in the following decreasing order : urea - formaldehyde, urea, ammonium chloride, ammonium sulphate, ammonium nitrate, ammonium sulphate plus organic matter.

648.
REDDY, M.V. (1981) Microarthropods and the rate of litter disappearance in a pine plantation ecosystem of North-Eastern India.

Pedobiologia 22 : 339-344.

The population structure of microarthropod cenosis associated with needle litter disappearance in a pine (Pinus kesiya) plantation was studied during May 1976 to June 1977. For this purpose 10 g quantities of the air-dried litter contained in nylon bags were placed at the forest site under a cover of freshly fallen litter.

The litter microarthropid population increased after June 1976, but decreased in August; a sudden increase again occurred in October to a level of 19.4% of the total faunal population. In February 1977, the population level had decreased to 1.8% of the total faunal population, but this was followed by a second peak in June (15.7%).

649.
RELWANI, L.L. (1963) Role of blue green algae on paddy yield. Curr. Sci. 32 : 417-418.

Blue-green slimy formations consisting of Nostoc, Anabaena, Microcystis, Phormidium and Aphonothece were scraped from a soil surface, air-dried and inoculated at the rate of 3.0 g/15 kg soil in pots or 340 g per plot 1/187.06 hectare in area. In pots the algae could provide an effective susbitute for NH_4-N; the yield response to inoculation was a little better in the presence of lime. Results in the field were similar.

650.
REMA DEVI, L. and SAM RAJ, J. (1966) The rhizosphere microflora of tomato as influenced by the age of the plant.

Agric. Res. J. Kerala 4 : 60-66.

A definite microbial population was established in the rhizosphere of three-day-old seedlings, and it showed a positive correlation with the age of plant up to maxima at the flowering stage. Actinomycetes and fungi increased for 20 days, with only minor changes afterwards, whereas the bacterial population steadily increased until the period of flowering.

651.
RENGASWAMI, G. (1956) The need for research on soil actinomycetes in India.

Madras agric. J. 43 : 253-259.

The article suggests needed research on soil actinomycetes which are (i) A thorough survey of the actinomycete population in different types of soils and at various depths, (ii) screening of actinomycetes for antibiotic production, vitamin production and for other industrial and agricultural utility, (iii) Study of the part played by actinomycetes in the decomposition of organic matter, humus formation and soil fertility in general, (iv) Study of the interrelationships of actinomycetes with other microorganisms in soil and to explore the possibilities of controlling the injurious microorganisms by favouring the processes brought about by the beneficial actinomycetes and (v) Identification and classification of actinomycetes and of the soil types in particular.

652.
RENU GROWER and KUNDU, B.S (1983) Growth and nitrogen fixation by bacterial isolates from barley roots.

Curr. Sci. 52 : 974-976.

Results of this study suggest that barley roots are associated with a high concentration of nitrogen fixing bacteria which remain active till flowering stage. These are mainly of two types, *Pseudomonas* sp and gram-negative rods. They need a mesophilic temperature, neutral pH, specific carbon and nitrogen compounds for optimal nitrogen fixation.

653.
REWARI, R.B.; SEN, A.N. and SEN, A. (1965) Nitrogen fixation by cluster bean (Cyamopsis tetragonoloba(L.)Taub.) in relation to phosphate uptake from soil.

Indian J. agric. Sci. 35 : 162-167.

In a 3-year pot experiment, the amount of N fixed by cluster bean plants was positively and significantly related to uptake of PO_4. Increase in N fixation with PO_4 uptake was higher in poor than in rich soil.

654.
REWARI, R.B., SEN, ABHISWAR and PANDEY, S.L. (1957) Excretion of nitrogen from the roots of guar bean (Cyamopsis psoraloides).

J. Indian Soc. Soil Sci. 5 : 237-245.

During a three year pot culture study of the nitrogen content of a soil with and without guar growing for about two and a half months, evidence was obtained of definite excretion of nitrogenous substances from the roots of the legume to the soil.

Excretion of nitrogen from the roots of guar and the nitrogen fixation by the legume both increased as a result of treatment of the soil with superphosphate. The amount of excretion varied from 8.1 to 20.9 per cent of the total nitrogen fixed by the legume.

In an average soil, the excretion in relation to the fixation of nitrogen is unaffected by treatment of the soil with superphosphate. In soil, poor in nitrogen, the percentage excretion decreased with the application of superphosphate.

655.
ROY, A.B. and MUKHERJEE, M.K. (1957) A new type of nitrogen fixing bacterium. Nature. 180 : 236.

A N-fixing bacterium showing certain morphological and physiological differences from known N-fixing organisms was isolated from soils cropped with jute. The bacterium utilizer carbohydrates especially glucose and mannitol, as energy material.

656.
ROY, A.B. and MUKHERJEE, M.K. (1961) Loss of ammonia during active nitrification from culture solution with different soils of West Bengal. Proc. nat. Acad. Sci. India 31A : 199-204.

The rate of nitrification in incubated samples was primarily affected by soil texture; it was highest in light to medium soils and was increased by high levels of exchangeable calcium. Loss of NH_3 was mainly caused by alkaline reaction and volatilization by evaporation, and was reduced during nitrification.

657.
ROY, M.B. (1927) Studied in intensive bacterial oxidation. II - The effect of various substances on the rate of nitrification. J. Indian Inst. Sci. 10A : 100-101.

The stimulating effect of peptone is due to its furnishing necessary food for the organisms and also the possibility of its decomposition products being capable of conversion into nitrates. The action of iron salt is doubtless catalytic. The ammonical liquor of gas works, if sufficiently diluted, could be oxidised directly to nitrate in presence of $CaCO_3$ without the necessity for the intermediate formation of ammonium sulphate.

658.
ROY MAHAPATRA S.S.; MISHRA, A.K. and CHANDRA, D. (1985) Solubilisation of rock phosphate by Thiobacillus ferrooxidans. Curr. Sci. 54 : 235-237.

Results show that the soil treatment with rock phosphate in combination with pyrite and T. ferrooxidans might show a way in the near future for better soil management. Authors have reported the involvement of iron-oxidizing chemoautotroph Thiobacillus ferrooxidans in solubilizing phosphorus from low grade rock phosphate mixed with pyrites.

659.
ROY, N.K. (1982) Effect of diphenyl dichloromethyl phosphonate, a newly developed fungicide on soil microflora.

Trans. 12th intern. Congr. Soil Sci., New Delhi 6 : 51.

By direct and indirect means a large portion of these pesticides whether applied in foliage or on soil is getting entry into the soil. Knowledge of metabolic fate of pesticides is essential for assessing the hazards associated with its residues in the soil. Studies were undertaken to evaluate the effect of a newly developed fungicide, diphenyl dichloromethyl phosphonate on microbial population, its rate of degradation and the nature of degraded products in soil. The results showed that the fungicide did not have any harmful effect on soil microbes even at concentration higher than the recommended dosage. The rate of degradation of the fungicide in soil was faster than on rice plant studied earlier. The half life of the fungicide was found to be 6 days on soil surface. The final product obtained is dichloromethyl phosphonic acid, a non-toxic product.

660.
ROY, R.Y. and GUJARATI, S. (1965) A new species of Dactylaria from soil. Lloydia 28 : 53-54.

Dactylaria fulva sp. nov. differs from previously described species in producing branched and proliferating conidiophores and hayline 3-celled cylindrical conidia.

661.
ROY, R.Y. and GUPTA, P.C. (1967) Phaeotrichoconis terrestre sp. nov., from soil.

Curr. Sci. 36 : 528-529.

This new species was **isolated** from the rhizosphere of Ocimum sanctum Linn.

662.
ROYCHOUDHURY, P.; DADARWAL, K.R. and VENKATARAMAN,

G.S. (1978) Antigenic and nodulation properties of Rhizobium trifolii transformants.

Curr. Sci. 47 : 457-458.

The antigenic homology of five Rhizobium trifolii nif+ transformants derived from Rhizobium trifolium RT/3 and Azotobacter chroococcum B_3 with strain RT/3 is confirmed. Nodulation tests on Trifolium alexandrinum variety tetraploid C further confirmed that the transformants were derived from strain RT/3 and maintained its symbiotic properties.

663.
ROYCHOUDHURY, P.; KRISHNAMURTI, G.S.R. and VENKATARAMAN, G.S. (1980) Effect of algal inoculation on soil aggregation in rice soils. Phykos 19 : 224 - 227.

The aggregation status show a remarkable increase after inoculation. Simple and complex polysaccharides liberated by algae were responsible for the icreased soil aggregation.

664.
RUPELA, O.P. and TAURO, P. (1973) Isolation and characterisation of Thiobacillus from alkali soils. Soil Biol. Biochem., 5 : 891-897.

Alkali soils are poor in sulphur-oxidizing bacteria. By appropriate enrichment, two bacterial strains resembling Thiobacillus thiooxidans and T. novellus possesses characteristics which make it suitable in reclamation of alkali soils by sulphur amendment.

665.
RUPELA, O.P and TAURO, P. (1973) Utilisation of Thiobacillus to reclaim alkali soils. Soil Biol. Biochem., 5 : 899-901.

In laboratory experiments addition of S to two alkali soils decreased the soil pH and exchangeable Na, and increased soluble Na and SO_4 contents; these changes were greater with soil inoculation with a strain of Thiobacillus novellus.

666.
RUPELA, O.P.; JOSEY, D.P.; TOOMSAN, B.; MITTAL, S.; DART, P.J. and THOMPSON, J.A. (1982) Application of inherent antibiotic resistance to ecological studies of rhizobia. In Biological Nitrogen Fixation Technology for Tropical Agriculture (edited by Graham, P.H. & Harris, S.C.). Centro Internacional de Agricultura Tropical (CIAT) Cali, Colombia. p 625-629.

In a study of 473 field isolates from a trial inoculated with a streptomycin (str 200) marked mutant, the use of 30 characteristics (10 antibiotics X 2-4 concentrations) classified the strains into 203 groups when all characteristics had to match perfectly, and into 119 groups when one mismatching characteristic was allowed. The 22 isolates having str 200 resistance were placed in three groups with no mismatches, and into two groups with one mismatching character allowed.

667.
SACHCHIDANANDA, J. and SWARUP, G. (1967) Additional nematophagous fungi from Delhi soils.

Curr. Sci. 36 : 677-678.

The morphology of Stylopage hadra Drechs., S. grandis Duddington, and Dactylaria brochopaga Drechs is described.

668.
SACHDEV, M.S. and LUTHRA, V.K. (1981) Effect of simazine and atrazine on the decomposition of 14C-glucose in soil.

J. Nucl. agric. Biol. 2 (10) : 53-56

The evolution of $14CO_2$ from 14C-glucose in an alluvial soil during 234 days decreased from 72% in the case of untreated soil to 50.5% and 17.5% following treatment with simazine and atrazine, respectively. The percentage of applied C in humic acid and humin fractions of soil OM was increased by the herbicides.

669.
SADANA, J.C. and KHAN, B.M. (1977) Nitrogen fixation.

J. Scient. Ind. Res. 36 : 510-533.

Recent advances in various aspects of biological fixation of N are reviewed and the aspects covered include: assay techniques, reactions catalyzed by nitrogenase, components of nitrogenase and cross-reaction between components, requirements for a unit of nitrogenase, mechanism of action of nitrogenase activity and biosynthesis, glutamine synthetase, glutamate synthase, glutamate dehydrogenase, molecular biology of N fixation, application of biological N fixation, microbial production of ammonium fertilizer from N_2, economics and limitations of biological N fixation, and inorganic organic complexes involved in N fixation.

670.
SADASIVAM, S. (1978) Nitrogen fixation by different isolates of Rhizobia in a defined medium. pp. 244-247. In S.P. Sen, Y.P. Abrol and S.K. Sinha (eds.), Nitrogen Assimiiation and Crop Productivity. Associated Publishing Company, New Delhi.

Changes in the total nitrogen, total protein and glutamic oxaloacetic transminase activity during the growth period of five different isolates of rhizobia in the CS 7 medium were studied. A direct relationship between the transminase activity and quality of nitrogen fixed as well as total protein assimilated by the cells was observed.

671.
SADASIVAN, T.S. (1939) Succession of fungi decomposing wheat straw in different soils, with special reference to Fusarium culmorum. Ann. Appl. Biol. 26 : 497-508.

Fusarium culmorum is an important primary colonizer of dead wheat straw buried in the soil and has to be considered as a soil inhabiting fungus.

672.
SAHA, K.C. and MANDAL, L.N. (1979) Distribution of nitrogen fixing blue-green algae in some rice soils of West Bengal.

J. Indian Soc. Soil Sci. 27 : 470 477.

In an incubation experiment with ten alluvial rice soils, of West Bengal, the potentiality of algal fixation of nitrogen was found to be 27.66 to 73.33 µg/g soil/week. The dominant genera in these soils were Nostoc, Anabaena and Cylindrosperumum followed by Scytonema, Calothrix and Mastigocladus. The magnitude of fixation of nitrogen in soil culture and in mineral nutrient culture was found to be different. Among the eight algal species isolated, three of Nostoc and one of Scytonema were found to be efficient nitrogen fixers, Nostoc muscorum being the most, in media either with or without combined nitrogen.

673.
SAHA, K.C. and MANDAL, L.N. (1980) Effect of inoculation with nitrogen-fixing blue-green algae on the population of total bacteria, Azotobacter and phosphate solubilizing bacteria in partial sterilized waterlogged rice soil.

J. Indian Soc. Soil Sci. 28 : 375-380.

The effect of inoculation with a mixed culture of two nitrogen fixing blue-green algae, Aulosira fertilissima and Nostoc muscorum on the growth and proliferation of total viable bacteria, Azotobacter and phosphate solubilizing bacteria were investigated in a partial sterilized alluvial waterlodded rice soil. The inoculation harboured increasing number of total viable bacteria, Azotobacter and phosphate solubilizing particularly during the growing period of inoculated algae. This was more pronounced when phosphate and molybdate were added to the soil both in presence or absence of rice plant.

674.
SAHA, K.C. and MANDAL, L.N. (1980) Fixation of nitrogen by blue-green algae in acidic and lateritic rice soils of West Bengal.

J. Indian Soc. Soil Sci. 28 : 98-102.

The fixation of nitrogen in some acidic and lateritic lowland rice soils of West Bengal due to inoculation of nitrogen fixing blue-green algae, Nostoc muscorum was investigated in a laboratory inoculation experiment. Inoculation without correcting the soil acidity could not bring about any increase in the total nitrogen content of the soils but the same after addition of lime was, however, found to cause a significant fixation in most of the soils. This was further encouraged by phosphorus application. The efficacy of molybdenum and the higher level of phosphorus application in increasing nitrogen fixation was found to vary from soil to soil.

675.
SAHA, S.K.; SHANKER, J. and DE, S.K. (1981) Effect of salts of fluoride on the decomposition of organic matter (wheat straw) in soil. J. Environ. Biol. 2 : 87-89.

Fluoride in general reduced the release of NH_3-N from the decomposition of carbohydrate organic matter in soil. This effect may be due to their direct effect on the enzyme responsible.

676.
SAHRAWAT, K.L. (1982) Comparative evaluation of karanjin and extracts of karanja (*Pongamia glabra* Vent.) and neem (*Azadirachta India* L.) seeds for retardation of nitrification of urea in soil. J. Indian Soc. Soil Sci. 30 : 156-159.

Ability of alcohol extracts of karanja and need seeds to retard nitrification of urea was compared with that of karanjin an incubation experiment. Treatment of urea with the extracts at the rate of 30% of the N rate was comparable to karanjin applied at 5 and 10 per cent concentration. The extracts were effective in retardation of nitrification upto 45 days while karanjin was effective upto 60 days. The different patterns of inhibition of nitrification observed with karanjin and neem seed extracts suggest that their mixture may be better material for retardation of nitrification in soil.

677.
SAHRAWAT, K.L. and PARMAR, B.S. (1975) Alcohol extract of neem (*Azadirachta indica* L.) seed as nitrification inhibitor.

J. Indian Soc. Soil Sci. 23 : 131-134.

Results indicated that the alcohol extract of neem effectively conserves ammonical nitrogen. The lipid associates of neem check the conversion of ammonium to nitrite and the subsequent nitrate. A remunerative outlet for these 10-12 per cent of the biologically active components of the seed can be of help in strengthening the economy of neem seed.

678.
SAKSENA, R.K. and SARBHOY, A.K. (1963) Ecology of the soil fungi of Uttar Pradesh. I. Fungi in different soils at Allahabad.

Proc. natn. Inst. Sci. India 29 B : 207-224.

A Gangetic alluvial soil and a black cotton soil were examined at different depths and seasons. 75 fungal species were identified; 24 species belonged to 19 Phycomyceta genera, and two Ascomycete genera were represented. 19 species of Aspergillus and eight species of Penicillium were recorded. Other organisms belonged to the fungi imperfecti. No Basidiomycetes were found.

679.
SAKSENA, R.K.; KRISHNA NAND, F.N.I. and SARBHOY, A.K. (1967) Ecology of the soil fungi of Uttar Pradesh. IV. Bundelkhand and Gangetic tract soils.

Proc. natn. Inst. Sci. India 33 B : 298-306.

Soils in these regions (35-42 in annual rainfall) are predominantly calcareous (up to 0.84%Ca) and slightly alkaline (pH7-8.4). Soil properties (moisture content, C and P contents, etc.) and frequency of microfungi at various soil depths during different seasons are tabulated. About 60 species of Phycomycetes, Ascomycetes and Deuteromycetes were identified, and a new species (Phoma terrestre) was isolated.

680.
SAKSENA, R.K.; NAND, K. and SARBHOY, A.K. (1967) Ecology of the soil fungi of Uttar Pradesh. II. Soils of the Himalayan

forests and their microfungi.

Proc. natn. Inst. Sci. India 33 : 144-153.

Samples were taken from red loam, brown forest, podzolic, meadow and maize-cultivated soils and from them 75 species of fungi, belonging to Phycomycetes, Ascomycetes and Deuteromycetes were isolated and identified. The maximum number of Penicillia was recorded in the rainy season, that of Aspergilli in the winter. In the winter and rainy seasons, maximum numbers of fungi occurred in the meadow and red soils and minimum in the brown forest soil. In the summer, maximum numbers were recorded in the red and meadow soils, less in the brown forest soil, and least in the maize and podzolic soils.

681.
SAKSENA, R.K.; SARBHOY, A.K. and NAND, K. (1967) Ecology of the soil fungi of Uttar Pradesh. III. Soils of the sub-Himalayan tract and their microfungi.

Proc. natn. Inst. Sci. India 33 : 154-161.

Forty-five species of fungi were isolated from forest soils of the sub-Himalayan tract and identified as belonging to Phycomycetes, Ascomycetes and Deuteromycetes. Few Penicillia occurred in the soils compared with Aspergilli. Maximum numbers of Fusaria were recorded during the summer in almost all soils.

682.
SANDHU, D.K.; SURINDER SINGH and WARAICH, M.K. (1980) Thermophilous fungi of decomposing sugarcane bagasse. Can. J. Bot. 58 : 2015-2016.

Twenty species of thermophilous fungi were isolated from sugarcane bagasse at Batala in Northern India. Seven species are reported from bagasse for the first time and Aspergillus caespitosus is a new record for India. Aspergillus fumigatus was the most common fungus followed by Mucor pusillus. The temperature responses of the fungi revealed 5 microthermophiles,

11 thermotolerant and 4 thermophiles. Three species of thermophilous fungi are reported in addition to those already recorded in the literature.

683.
SANDHU, K.S.; RANDHAWA, K.S. and CHAHAL, V.P.S. (1978) The effect of some herbicides on the rhizosphere microflora in the seed crop of okra (Abelmoschus (= Hibiscus) esculentus (L.) Moench).

Indian J. Ecol. 5 : 61-66.

Field experiments were conducted on loamy sand soil at the Punjab Agricultural University, Ludhiana during 1974-75 and 1975-76. The bacterial population was higher in three controls, i.e. control weeded, left unweeded and handpulled, than that in all herbicidal treatments. Alachlor at 1.94 kg a.i./ha resulted in a significant increase in the actinomycete population, whereas alachlor at 2.5 kg. EPTC + alachlor at 3.75 + 1.25 kg and fluchloralin at 1.2 and 0.9 kg/ha were on a par with the controls, handpulled and left unweeded. The fungal population was relatively unaffected by the various herbicidal treatments, except nitrofen + linuron at 1.25 + 0.25 kg EPTC at 0.75 kg, dichlormate at 5 kg. EPTC + nitrofen 3.75 + 0.62 kg and nitrofen at 1.25 kg/ha, which caused a significant reduction in the fungal population as compared with that in the controls.

684.
SANKARAM, A. (1968) Microbiological aspects of soil fertility with special reference to rice culture.

Indian J. agric. Chem. 1 : 7-11.

The beneficial effects of inoculating rice soil with blue-green algae (and inoculating legumes in the rotation with Rhizobium) are due mainly, though not exclusively, to the fixation of atmospheric N.

685.
SANKARAM, A. (1971) Work done on blue-green algae in rela-

tion to agriculture.

Tech. Bull. (Agric.) Indian Coun. Agric. Res. 27 : 28.

A review of algae in soil, algal physiology and nutrition, production of extra-cellular substances, response to pesticides and preparation and use of algal inoculants.

686.
SANKARAM, A. and SUNDARA RAO, W.V.B. (1962) Synthesis of vitamin B_{12} by soil bacteria. Curr. Sci. 31 : 334-336.

Bacteria synthesising vitamin B_{12} occurred in the rhizosphere of berseem, cowpea and wheat, rich synthesisers () 15 mug/ml) occurring mainly in berseem and poor synthesisers in wheat. Manuring promoted bacterial synthesis of the vitamin.

687.
SANKARAM, A.; RAJU, P.V.L.N. and KAILASH, S. (1963) Influence of phosphate on nitrogen fixation by Rhizobium species. Indian J. Microbiol 3(1) : 17-20.

P was applied at 15, 30 and 45 kg/ha P_2O_5 to sunhemp growing in unsterilized black soil and in sterile sand. 15 kg/ha increased the total protein content of plants in sterile sand. In the black soil, P did not affect total protein content.

688.
SANKARAN, K.V.; ZACHARIAH, S. and LEELAVATHY, K.M. (1980) Aspergillus brunneo-uniseriatus var. nanus from soil. Mycologia 72 : 625-629.

A new variety of Aspergillus brunneo-uniseriatus is described and figured. It differs from A. brunneo-uniseriatus Singh & Bakshi var. brunneo-uniseriatus mainly in less flocculent growth and smaller conidial heads, conidiophores and conidia. The fungus was isolated from soil and rhizosphere of Piper nigrum L. from the Calicut University campus on a few occasions in 1978.

689.
SANKHLA, H.C.; MASIH, B. and MATHUR, R.L. (1969) Effect of oilcakes extract on the growth of three soil fungi.

Indian Phytopath. 22 : 271-273.

The effects of extracts from mustard, castor, Sesamum, groundnut and linseed cakes on Rhizoctonia bataticola (Taub.) Butler, Fusarium oxysporum, F. coriandertii and F. oxysporum, F. cumini were studied in culture medium. Optimal concentrations of oilcakes extracts varied with the type of cake and the fungus.

690.
SANORIA, C.L.; SINGH, K.L.; RAMAMURTHY, K. and MAURYA, B.R. (1982) Field trials with Azospirilum brasilense in an Indo-Gangetic alluvium. J. Indian Soc. Soil Sci. 30 : 208-209.

There were good increases in yields due to Azospirillum alone over the control. The plant height was affected significantly in case of paddy and nearly significantly in case of sorghum and bajra. Residual effect of Azospirilum alone was clearly revealed by increase of wheat grains in rabi season. Use of Azospirillum along with no basal application of nitrogen was found to be more desirable than applying 30 kg N/ha.

691.
SANYASI RAJU, M. and VARADARAJAN, S. (1955) The effect of irrigation on the microbiological changes of the rainfed black soils of Rayalaseema.

J. Indian Soc. Soil Sci. 3 : 77-82

Nine years' study of the microbiological changes brought about when rainfed, black soils of Rayalseema (Siruguppa) were converted into irrigated lands reveals that the microbiological population increased considerably during the first year of irrigation. It was followed by a fall in the subsequent years and got stabilised. Maximum values were observed, in general, in block VIII which was under garden land cultivation.

The microbial activity was influenced mainly by the supply of available organic matter. Here also, Block VIII was the best of all the treatments. The essential factors for maintaining high microbial population and activity are (i) irrigation; (ii) available organic matter; and (iii) aeration.

692.
SARKAR, K. and VENKATA RAO, B.V. (1966) Influence of soil phosphorus status on nitrification in red soils of Mysore.

J. Indian Soc. Soil Sci. 14 : 97-100.

A laboratory study on the rate of nitrification in red soils of Mysore wherein phosphorus status built up by repeated applications of increasing levels of phosphate indicated that the process of nitrification was not dependent on the soil phosphorus status.

693.
SARMA, K.S.B.; LAKSHMI-KUMARI, M; APTE, R. and RAO, N.S.S. (1973) Some physiological characteristics of Rhizobium meliloti and R. trifolii in relation to efficiency of symbiosis with lucerne and Egyptian clover.

Pl. Soil. 38 : 299-305.

No positive or significant correlations were found between symbiotic efficiency and the physiological properties of bacteria, but significant inverse relationships occurred between the formation of Salkowski positive compounds from tryptophane, and indole acetic acid destruction (in R. meliloti and R. trifolii) between NO_2 formation and NO_2 utilization (in R. trifolii) and between symbiotic efficiency and NO_2 utilization (in R. meliloti).

694.
SARMA, V.A.K. and DAKSHINAMURTI, C. (1971) Relative efficiency of bacterial gum and organic matter extract as aggregating agents in sandy - clay mixture.

Curr. Sci. 40 : 35-36.

The binding strength of organic extract from partly-decomposed leaf material has been compared with that of bacterial gum using the technique of displacement of liquid through porous beds. The flow rate was lowest with the aggregates bound by the bacterial gum. The organic matter extract and the mixture of organic matter and bacterial gum appear to bind the clay and sand particles into comparatively large-sized aggregates giving higher rates of flow.

695.
SASIDHAR, V.K. (1976) Bacterial fertilizers - A new approach in plant nutrition. Seeds & Farms 2(9) : 11-12.

The article highlights the importance of bacterial fertilizers in augmenting nitrogen fixation through free living and simbiotic micro-organisms specially Azotobacter, Rhizobia, blue-green algae etc. Bacterial fertilizers also play significant role in improving the solubility and availability of phosphorus.

696.
SATISHCHANDRA, K.M.; HIREMATH, R.V. and HEGDE, R.K. (1979) Assaying inoculum density of Rhizoctonia bataticola causing root rot of bean, in soil by using seeds.

Indian Phytopath. 32 : 145-146.

The percentage of seeds colonized was maximum in beetroot followed by okra. Maximum baiting by beetroot seeds was about 90 per cent with 24 per cent inoculum. Even with 4 per cent inoculum colonization was fairly substantial. Moderate colonization was obtained in the seeds of cereals viz. maize and wheat. Seeds of 3 leguminous plants tested at higher inoculum levels, the baiting by leguminous seeds was poor and the increase in colonization was not proportional to the increase in inoculum level.

697.
SATISHCHANDRA, K.M.; HIREMATH, R.V. and HEGDE, R.K.

(1980) Factors affecting the survival of Rhizoctonia bataticola in black soil. Pl. Soil 54 : 307-312.

The survival of Rhizoctonia bataticola in a black soil was adversely affected by high soil moisture levels. The half life decreased from 49 days at 20% of moisture-holding-capacity (MHC) to 13 days at 80% of MHC. Variations in soil pH between 5.1 and 7.9 and addition of Na_2SO_4 $MgSO_4$ $CaCl_2$ and NaCl did not affect fungal activity.

698.
SATTAR, M.A. (1975) Ammonification and nitrification of urea-N in soils.

J. Indian Soc. Soil Sci. 23 : 416-419.

Application of urea to noncalcareous dark grey floodplain, grey floodplain and noncalcareous brown floodplain, and grey floodplain soils increased the NH_4+ upto two weeks of incubation and to brown hill soils upto first week and decreased thereafter. The NO_3- concentration of the soils increased with increasing incubation time. The NH_4+ and NO_3- of the soils increased with increasing rates of applied urea-N.

699.
SATYANARAYANA, K.V.S. and DATTA, S.C. (1953) Sulphate reducing organisms in the soils of Rann area in Kutch. Curr. Sci., 22 : 145.

Sulphate reducing organisms were present in all the profiles but their activity was not uniform in the different horizons of the same profiles. A high activity was noticed in 7.5 to 37.5 cm layer of soils from the Rann area. In the deep gley horizon (165 to 177.5 cm) intense activity was noticed while in a profile with a surface gley horizon (30 to 80 cm) the activity was negligible. Gypsum bearing horizons showed high activity in some profiles and negligible in others. Similarly, in the iron accumulation layers also the activity was not consistently of the same order.

700.
SATYAPRASAD, K. and RAMARAO, P. (1983) Saprophytic colonization of Fusarium oxysporum f. sp. ciceri in soil. Folia Microbiol. 28 : 173-178.

The competitive saprophytic colonization (CSC) of soyabean and chickpea stem pieces by Fusarium oxysporum f. sp. ciceri increased with increasing inoculum density in inoculum/soil mixtures. Progressive dilution of autoclaved soils with unsterilized soil decreased the CSC. Lower temperatures favoured the colonization in both red sandy loam and black soils. Maximum colonization occurred at 40 ^{o}C indicating an inverse relationship between colonization and temperature.

701.
SAXENA, M.C.; SINGH, H.P. and VERMA, R.S. (1978). Symbiotic nitrogen fixation in soyabean under humid sub-tropical conditions of Pantnagar, pp.44-45. In S.P. Sen, Y.P. Abrol and S.K. Sinha (eds.), Nitrogen Assimilation and Crop Productivity. Associated Publishing Company, New Delhi.

Estimate of symbiotic nitrogen fixation in a late crop of soyabean (Glycine max. (L.) Merr.) was made using nodulating and non-nodulating isolines of the cultivar Lee grown in the field with 20 and 160 kg ha-1 of fertilizer nitrogen labelled with ^{15}N and adopting 'A' value technique. The nodulated crop yielded 25 quintals of beans per hectare when raised with 20 kg per hectare of starter fertilizer nitrogen and had a symbiotic nitrogen fixation of about 114 kg N ha-1. Application of fertilizer nitrogen @ 160 kg N ha-1 decreased the symbiotic N fixation, the magnitude of decrease being higher when the application was done in 3 splits in comparison to whole application at planting.

702.
SEN, A. (1955) Studies on nitrogen fixation by Azotobacter occurring in tank silts. Indian J. agric. Sci. 25 : 131-142.

Morphological, cultural and biochemical characteristics are described of pure strains of Azotobacter isolated from different air dried tank silts. The average N fixation during a period of two weeks varied between 2 - 7.4 mg/g of mannit with pure strains and 4.4 - 13.7 mg with a mixed flora.

703.
SEN, A. (1956) Nitrogen fixation by algal associations in tank silts. Indian J. agric. Sci. 26 : 89-93.

Tank silts inoculated into a sterile N-free nutrient solutions gave rise to algal growth. Algal scrapings transferred to N-free nutrient solutions fixed 0.3-5.0 mg N/100 CC Benecke's solution and organisms, including azotobacter, associated with the algae (grown on Ashby's mannite solution) fixed 0.8 - 6.7 mg, all during a period of two months. Algal growths harbouring N-fixing organisms in tank silts may be responsible for the N fixation in soil treated with silts.

704.
SEN, A. (1963) A note on a heavily capsulated alkaliphil bacillus from soil. Indian J. agric. Sci. 33 : 139-144.

A bacillus similar to Bacillus krzemienieski and B. gommeaux, but gram positive, was isolated from an alkaline soil. Two strains were recognized, both growing best at pH 9.0. The bacillus grew in normal and alkaline soils without lowering the N content. It increased the pH of the normal soil and decreased that of the alkaline soil. In both soils, more particularly the alkaline one, the bacillus appreciably increased aggregation.

705.
SEN, A. and BAROOAH, P.P. (1964) Association of Azotobacter and Beijerinckia in Indian Soils. Bull. Nat. Inst. Sci. India 26 : 227-237.

Beijerinckia was observed outside as well as in the tropics and in soils other than laterites ; contrary to previous belief it was found in nearly 70% of soils pH < 6.0. In about 26% of soils examined, Azotobacter and Beijerinckia occurred together and antagonism was marked in such associations. Reaction of the substrate decide dominance ; most strains of *Beijerinckia* showed maximum N fixation and predominated over *Azotobacter* at low pH levels, but at pH 7.0 *Azotobacter* dominated.

706.
SEN, A. and ISWARAN, V. (1959) Variation of characteristics and nitrogen-fixing capacities of *Azotobacter* in some Indian soils. *Soil Sci.* 87 : 46-49.

Of the 40 strains isolated, 11 showed tolerance to pH 5 and 17 to pH 10, while 8 grew between pH 5 to 10 and 20 were intolerant both to acid and alkaline conditions ; only one strain grew on a medium of salinity of 3% NaCl.

707.
SEN, A. and PAUL, N.B. (1957) Solubilisation of phosphates by some common soil bacteria. *Curr. Sci.* 26 : 222.

In cultures containing $(NH_4)_2SO_4$ as the source of N, *B. subilis, B. mycoides, B. megatherium* and *B. mesentericus* released soluble P from Ca and Fe phosphates, Caglycophosphate and lecithin. *B. mesentericus* and B. *mycoides* did not release soluble P from bonemeal but utilized the soluble P already present. With urea as the N source *B. subtilis* utilized soluble P in Ca and Fe phosphates, but the other bacteria liberated soluble P therefrom. *B. megatherium* liberated soluble P from lecithin, but the other bacteria utilized large amounts of the soluble P in lecithin.

708.
SEN, A. and PAUL, N.B. (1958) Occurrence of phosphobacterium in the glands of *Cassia occidentalis. Indian J. agric. Sci.* 28 : 21-29.

The organism isolated from the small red glands attached to the petioles of the non-nodulating legume *Cassia occidentalis* developed freely in presence of combined N and showed characteristics of phosphobacteria described by Russian authors. The presence of these organisms in soil is likely to increase P availability from ferric phosphates in lateritic soils and from other phosphates like bone meal, provided N fertilizers such as ammonium sulphate and urea are applied.

709.
SEN, A.N. (1966) Inoculation of legumes as influenced by soil and climatic conditions.

Indian J. agric. Sci. 36 : 1-7.

Most effective strains of Rhizobium of Cincer arietinum were isolated from nodules that had the highest lethemoglobin content. Strains isolated from plants grown at different places showed great variations in some biochemical characteristics and in ability to fix atmospheric N_2. An effective Rhizobium strain should only be used at places that have soil and climatic conditions similar to those of the original habitat.

710.
SEN, A.N.; REWARI, R.B. and SEN, A. (1966) Effect of humus from tank silt on nitrogen fixation by Azotobacter.

Indian J. agric. Sci. 36 : 21-24.

In the humus fraction from 19 samples of tank silts (ranging from 0.27 to 0.85%), the C content was remarkably constant but the N, P and Fe contents showed wide variations. Unlike humus formed during aerobic decomposition of organic matter, the humus formed through anaerobic decomposition under water depressed N fixation by Azotobacter chroococcum even if present in small amounts in the growth medium; the degree of depression showed a significant inverse relationship to the N content of the humus.

711.
SEN, ABHISWAR and SEN, A.N. (1956) Survival of Rhizobium japonicum in stored air dry soils.

J. Indian Soc. Soil Sci. 4 : 215-220.

Out of nine cultivated soils which were previously observed

to contain Rhizobium japonicum and which were stored in dry air condition for over nineteen years, the organism had been observed in the two soils from Akola and Jorhat. The organism from these two soils was almost similar in size and in other morphological, cultural and physiological characteristics. The organism from Akola was fast growing while that from Jorhat was a slow growing one. The organism from both the soils infected soyabean equally well.

The longevity of the organism in these two soils could not be attributed to the presence of capsules as it was found, in both the soils, to be non-capsulated. It could neither be attributed to any soil constitutent or property which differentiated these two soils from others.

The results of the present investigations do not offer any explanation for the longevity of the organism in Akola and Jorhat soils and the absence of the same in others but suggest that the survival of the organism in the two soils may be due, as had been observed by some workers, to the presence in the soil solution, of substance of organic origin or of trace elements, which may have some protective influence on the rhizobium against death by desiccation.

712.
SEN ABHISWAR, MEHTA, P.M. and REWARI, R.B. (1960) Relationship of the values of available phosphate as determined by chemical methods with nitrogen fixation in soils by Azotobacter.

J. Indian Soc. Soil Sci. 8 : 65-69.

During the present investigations, 19 soils from different parts of India were analysed for their contents of total and available P_2O_5. The available P_2O_5 in the soils was determined by extraction with 1 per cent citric acid, 1 per cent potassium carbonate and 0.5 M sodium bicarbonate. The investigation included also determination of nitrogen fixation by a strain of Azotobacter in the soils by Remy's method.

Nitrogen fixation in soils was found to be positively and significantly correlated with the total phosphate and also with the available phosphate content as determined by the method of extraction with 1 per cent solution of citric acid only. It did not show any correlation with the available P_2O_5 content in the soils as determined by extraction with alkaline solutions.

713.

SEN GUPTA, B.; SEN, M. and SEN, S.P. (1978) Biological nitrogen fixation and nitrogen nutrition of crop plants. pp.56-70. In S.P. Sen, Y.P. Abrol and S.K. Sinha (eds.), Nitrogen Assimilation and Crop Productivity. Associated Publishing Company, New Delhi.

Biological N_2-fixation is an essentially endergonic process. In photosynthetic bacteria and blue green algae, the process is coupled to photosynthetic energy supply and thus, does not need the addition of organic material essential for heterotrophic N_2-fixers. In symbiotic N_2-fixation the energy is supplied by the green plants. Considerable organic matter is also leached out from the leaf surface, which harbours a variety of microorganisms including N_2-fixers and the available organic material can be used for N_2-fixation. A large number of such N_2-fixing microorganisms have been isolated from the leaf surface of a variety of plants; these microorganisms reduce acetylene to ethylene and when sprayed on crop plants in N-less sand culture or on soil, improve the growth, N-content and the yield of such plants markedly. In the case of IR-8 and IR-26 rice and Sonalike wheat the yield in quintals ha-1 was 67.62 and 25, respectively. Other crop plants which have so far been found to increase similarly in growth and yield include jute, potato, soyabean and several vegetables. *Rhizobium* when sprayed on the foliage of soyabean plants in N-less sand culture and on soil improves growth as remarkably as in the seed-inoculated ones. It is suggested that the large number of N_2-fixing microorganisms which occur in nature can be judiciously exploited for the nutrition of crop plants, reducing or eliminating the need of application of nitrogenous fertilizers. Such plasmid-curing agents as acridine orange and ethidium bromide eliminate the N_2-fixing capacity and the effects are not mutagenic. The *nif* gene thus, may be plasmid-carried and the introduction of such genes to higher plant cells by appropriate genetic engineering methods should be possible.

714.
SEN, J. (1929) Is bacterial association a factor in nitrogen assimilation by rice plants? Agric. J. India 24 : 229-231.

The contents of rice roots when examined under the microscope showed the presence of bacteria. Culture solutions containing a sugar and minerals but no nitrogen in any form when inoculated with the contents of the rice roots showed slimy bacterial growths which was further favoured by anaerobic conditions, clearly indicate assimilation of nitrogen from the gaseous atmosphere.

715.
SEN, S.P. (1982) Energy recycling in nitrogen fixation : Problems and prospects. In Frontiers of Research in Agriculture (edited by Roy, S.K.). Indian Statistical Institute, Calcutta, p 204-214.

The efficiency is maximum in the photosynthetic bacteria and cyanobacteria, followed by endosymbiotic associations. The advantages and limitations of phyllospheres and rhizosphere N_2-fixation are discussed.

716.
SEN, S.P. (1983) Biological nitrogen fixation : some recent developments and unanswered questions.

Proc. Natl. Symp. Adv. Front. Pl. Sci. Jodhpur 9 : 96-98.

The article emphasises that the efficiency of the coupling of photosynthesis and nitrogen fixation as an important determinant of life processes in the biosphere. The article also stresses possibilities for introduction of the nif gene into eukaryotic cells through genetic engineering.

717.
SESHACHARYULU, E.V. (1937) Nitrogen fixation and Azotobacter count on application of carbohydrates and other energy rich

materials to the soil, Part III.

Proc. nat. Acad. Sci. India 7 : 149-157.

Azotobacter counts were made in soils treated with carbohydrates and exposed to sunlight or kept in the dark. N fixation was doubled under light treatment whilst Azotobacter were 10 times less numerous without light. It is concluded that both light and energy materials assist nitrification.

718.
SESHMUKH, S.K. and AGRAWAL, S.C. (1983) Prevalence of dermatophytes and other keratinophilic fungi in soils of Madhya Pradesh (India). Mykosen 26 : 574-577.

Of 240 soil samples examined by hair baiting, 142 (59.16%) yielded dermatophytes and other keratinophilic fungi.

719.
SETHI, R.P. and BHATTACHARYA, A. (1978) Biochemical studies on Rhizobium trifolii.

Proc. Indian natn. Sci. Acad. B 44 : 71-77.

In nutrient medium, utilization of asparagine was greater by four foreign strains of Rhizobium trifolii than it was by two local Ludhiana strains. None of the strains reduced TTC, and only the Australian strain failed to reduce KNO_2. Chloroterracycline was the most inhibitory of the antibiotics tested, and penicillin was not inhibitory. In a pot experiment with and without surface sterilization of berseem seed, inoculation with the foreign strains increased nodule weight, increased the fresh weight, dry weight and the protein content of the plants more. In a field experiment, three of the foreign strains increased berseem yield.

720.
SETHUNATHAN, N. (1970) Foliar sprays of growth regulators and rhizosphere effect in Cajanus cajan Millsp. I. Quantitative changes. II. Qualitative changes in the rhizosphere and certain

metabolic changes in the plant.

Pl. Soil 33 : 62-70.

Sprays of naphthalene acetic acid (NAA) increased fungi, and those of 2, 4-D significantly increased bacterial populations in the rhizosphere, irrespective of concentration and plant age. Gibberellin and maleic-hydrazide (MH) sprays generally reduced microbial numbers in the rhizosphere. Occurrence of physiological groups in the rhizosphere, after spraying, followed the order; starch hydrolysers > nitrifiers denitrifers. Spraying with NAA and IAA, shifted the balance in favour of denitrifiers and increased Azotobacter in the rhizosphere. Starch hydrolysers in the rhizosphere were promoted by 2, 4-D, NAA and IAA, and depressed by gibberellin and MH; sprays of 2, 4-D and NAA stimulated amino-acid requiring bacteria, while MH tended to stimulate preferentially organisms that require complex nutrients.

721.
SETTY, R.A.; BALIGAR, C.C. and PATIL, S.V. (1970) Effect of altrazine on the rate of nitrification in black clay loam soil.

Mysore J. agric. Sci. 4 : 111-113.

When applied to the soil in conjunction with $(NH_4)_2SO_4$ atrazine at 10 ppm gave a higher nitrification rate after 2 weeks incubation than the control soil or soil receiving higher rates of atrazine did, but after 5 weeks incubation the nitrification rate was lowest in soil receiving the 10 ppm level.

722.
SHANTARAM, M.V. and RANGASWAMI, G. (1967) A comparison of the rhizosphere microflora of mixed crops.

Mysore J. agric. Sci. 1 : 7-12.

Mixed cropping of sorghum and sunnhemp increased the rhizosphere bacterial population of sorghum but decreased that of sunnhemp, compared with populations for the crops grown individually. The fungal populations of both decreased.

723.
SHANTARAM, M.V.; BALASUBRAMANIAN, A. and RANGASWAMY, G. (1973) Sequence of soil microflora decomposing maize straw and stubble added to a red sandy loam soil.

Madras agric. J. 60 : 1074-1079.

Among soil micro-organisms developing on maize straw and stubble incorporated into a red sandy loam, bacteria predominated over actinomycetes and fungi. In samples incubated at 50°C bacterial population was lower than in samples incubated at room temperature. Fungal populations were also drastically reduced in soils incubated at 50°C, but populations of actinomycetes did not vary to any significant extent.

724.
SHANTHAKUMARI, P. and SINHA, S.K. (1974) Nitrate reductase in relation to nodulation in cowpeas and mung beans.

Curr. Sci. 43 : 545-547.

High nitrate reductase activity was observed in cotyledonous leaves of both mung beans and cowpeas and there was a rhythmic behaviour in both plants. Root nodules appeared in both cowpeas and mung beans by the second week after sowing and reached a maximum in six and five weeks respectively for cowpea and mung beans. The entire N fixed by nodules in reduced form may not be used as such but converted to nitrate before utilization by the plants.

725.
SHARAWAT, K.L. (1980) Control of urea hydrolysis and nitrification in soil by chemicals-prospects and problems. Pl. Soil. 57 : 335-352.

A review is presented of recent work on the regulation of urea hydrolysis and nitrification in soils by employing chemical inhibitors. The benefits from control of nitrogen transformations, in terms of enhancing fertilizer nitrogen efficiency for crop production, and the problems associated with N metabolism of plants are discussed. Prospects of using cheap and effective indigenous materials and chemicals for control of urea hydrolysis and nitrification under specific soil situations are encouraging. Urease inhibitors may reduce problems associated with ammonia volatilization if this is not offset by leaching of urea. On the other hand retardation of nitrification appears useful in reducing losses due to leaching and denitrification, and may be more effective in improving the utilization of fertilizer N by plants able to metabolise NH_4-N and NO_3-N equally well.

726.
SHARAWAT, K.L. (1980) Nitrogen mineralization in acid sulfate soils. Pl. Soil 57 : 143-146.

Nitrogen mineralized in two acid sulfate soils under anaerobic incubation at 30°C for 2 weeks accumulated entirely as NH_4^+. Nitrification did not occur in either soil under conditions that stimulate nitrification. Thus denitrification may not be a serious problem in these soils.

727.
SHARMA, B.M. and GUPTA, R.S. (1983) Effect of algal application on rice yield in Jammu Division.

Phykos 22 : 176-179.

One five year experimental and thirteen demonstration trials at six locations during 1978-82, in Jammu division showed that algal contribution in terms of crop yield was equivalent to the application of 20 kg N/ha.

728.
SHARMA, C.R. and VERMA, J. (1979) Performance of lignite based carriers on the survival of rhizobia.

Sci. Cult. 45 : 393-394.

The survival of Rhizobium leguminosarum, R. trifolii and R. meliloti cultured in lignite with 10% addition of lucerne hay meal was 3-fold higher than that of the rhizobia cultured in lignite alone. Inoculation of seeds of pea, berseem and lucerne with rhizobia cultured in lignite/lucerne hay meal mixture increased nodulation and plant dry wt.

729.
SHARMA, D.D. and SINHA, S.K. (1980) A survey of antagonistic actinomycetes from the soils of Agra region.

Indian Phytopath. 33 : 581-585.

Out of 118 strains of actinomycetes isolated from Agra soils, 25.42 per cent were found to be antagonistic to both Colletotrichum gloeosporioides and C. falcatum, 24.57% to C. gloeosporioides only, 5.08 per cent to C. falcatum only and 44.91 per cent showed no antagonistic activity. The number of antagonists was higher in soils slightly alkaline in reaction and rich in organic matter. Two actinomycete isolates possessing strong antagonistic activity were further tested by the streak method against 41 micro-organisms.

730.
SHARMA, D.N. (1969) Influence of blue-green algae on the yield of paddy crop in presence and absence of organic matter and phosphate in field.

Indian J. agric. Chem. 2 : 22-26.

Yield increases on paddy soil inoculated with Anabaena naviculoides or Chlorella vulgaris were very small compared to those obtained by a mixture of organic matter (straw) and phosphate (basic slag).

731.
SHARMA, H.C.; LAL, F. and LAL, P. (1980) Effect of Rhizobium inoculation on pea irrigated with waters of different residual sodium carbonate levels.

J. Indian Soc. Soil Sci. 28 : 469-473.

Two field experiments were conducted on alkaline loamy sand soils to study the effect of Rhizobium inoculations (with and without Mo under pelleting and non-pelleting conditions) on pea irrigated with waters of different residual sodium carbonate (RSC) levels (3, 6 and 9 me/l). Rhizobium inoculation (non-pelleted) proved to be the best of all the treatments for pod and straw yields in both the years of study and also increased the content of total N in the soil significantly. The ECe, hydraulic conductivity, and total N in the soil and pod and straw yield decreased and pH and ESP values of soils increased with increasing levels of RSC in water in both the years.

732.
SHARMA, J.P. and SAXENA, S.N. (1982) Dynamics of soil phosphorus in the rhizosphere of Pisum sativum.

J. Indian Soc. Soil Sci. 30 : 230-231.

Rhizobium inoculation alone could not increase the Olsend's P, whereas pelleting of inoculated seeds with gypsum showed a significant increase.

733.
SHARMA, J.P. and SAXENA, S.N. (1984) Efficiency of rhizobial inoculation under phosphorus fertilization J. Indian Soc. Soil Sci. 32 : 499-500.

Both Rhizobial inoculation and pelleting of seed were found to increase the grain yield and P uptake. The increase in grain yield due to Rhizobial inoculation was three times over uninoculated control due to phosphate solubilization and nitrogen fixation.

734.
SHARMA, L.C.; SADHANA SAXENA and REDDY, B.N. (1983) Seed treatment of gram with fungicide, insecticide and Rhizobial inoculation. Andhra agric. J. 31 : 39-41.

Studies on compatibility and order of treatment insecticides (Aldrin 30 EC and Aldrin 5% dust), fungicide (Bavistin) and rhizobial seed inoculation as pre-sowing treatments separately or in combinations, one after the other indicated maximum nodulation and highest grain yield when Rhizobium inoculated seeds sown and Aldrin 5% dust applied in soil.

735.
SHARMA, L.N. and SAXENA, S.N. (1974) Influence of 2, 4-D on soil microorganisms with special reference to Azotobacter. J. Indian Soc. Soil Sci. 22 : 168-171.

Influence of sodium salt of 2, 4-D (2, 4-Dichlorophenoxy acetic acid) on multiplication of soil microorganisms was evaluated after inoculating the soil with Azotobacter chroococcum strain-41. Increasing concentration of 2, 4-D in soil upto 4.5 ppm level increased the counts of the fungi, Azotobacter and total bacteria up to 2nd, 14th and 32nd day after treatments, respectively, while the same decreased the counts of actinomycetes at all the levels of its application up to 20th day after the treatments. Application of 2, 4-D beyond 4.5 ppm level showed toxicity even to fungi, Azotobacter and total bacteria. Association of increase in Azotobacter and total bacteria counts with decrease in actinomycetes has been discussed.

736.
SHARMA, N,; SRIVASTAVA, L.L. and MISHRA, B. (1983) Studies on microbial changes in soil as a result of continuous application of fertilizers, farmyard manure and lime. J. Indian Soc. Soil Sci. 31 : 202-206.

Effect of continuous application of fertilizers, FYM and lime on microbial population of an acidic red loam soil was studied. While bacteria, actinomycetes, Azotobacter, cellulose decomposers and phosphate solubilizers increased in FYM, FYM+lime+PK and lime+NPK, fungi and Beijerinckia proliferated under N, NP and NPK treatments. Azotobacter was absent in the latter three due to unfavourable pH condition created. Microbial population in the wheat rehizosphere was invariably higher than that in the non-rhizosphere but FYM application alone or in combination with the inorganics and lime supported greater population of bacteria and actinomycetes. Ammonium sulphate incorporation stimulated fungal population both in the rhizosphere and non-rhizosphere. A significant positive correlation between the rhizosphere microflora and CO_2 evolution was observed.

737.
SHARMA, R.D. and SINGH, R.S. (1978) Survival of Fusarium moniliforme in soil.

Indian Phytopath. 31 : 352-356.

In the absence of native soil microflora in autoclaved soil hyphal propagules of Fusarium moniliforme remained viable for 382 days, while conidia were recovered only up to 215 days. When an autoclaved mixture of soil and maize debris was inoculated with the fungus, its development and survival through hyphal propagules was better than in residue free autoclaved soil for 259 days. Pre-colonised maize debris showed high fungal activity for 7 months in autoclaved soil, but in unautoclaved soil the activity declined within 4 months. The fungus did not produce chlamydospores in culture media, in soil or in host debris. Thickened hyphae were detected in pre-colonised straw buried in the soil. Mycelial mats buried in soil produced black sclerotial structures.

738.
SHARMA, S.; JAIN, N. and SHAH, A.K. (1985) Occurrence of Azotobacter chroococcum in Pothos scandens.

Curr. Sci. 54 : 142-143.

This paper reports the intracellular detection of Azotobacter chroococcum, a free living nitrogen fixer in Pothos scandens, commonly known as the money plant which can grow in plain water.

739.
SHARMA, S.S and BHATTACHARYA, A.K. (1958) Influence of

calcium phosphate incorporated with energy-rich material on nitrogen fixation. Proc. nat. Acad. Sci. India 27 : 265-267.

In alkali soil the presence of Ca phosphate alone or with glucose or starch promoted the oxidation of organic matter and increased N fixation.

740.
SHARMA, S.S. and BHATTACHARYA, A.K. (1961) Nitrogen fixation in soil in presence of potassium phosphate and organic matter. Proc. nat. Acad. Sci. India 31A : 118-123.

The fixation of atmospheric nitrogen by additions of dried powdered neemleaf or cowdung to alkaline soil was greatly enhanced by K_2HPO_4.

741.
SHAW, N.K. and DAS, A.C. (1979) Hydrolysis and utilisation of urea by soil fungi. I. Aspergilli and Penicillia.
Fertil. Technol. 16 : 104-106.

Five species of Aspergillus and 5 species of Ponicillium isolated from paddy fields were tested for their ability to hydrolyse and utilise urea supplied as a sole source of nitrogen. All the species tested were capable of hydrolysing urea to ammonia but the rapidity of such activity varied quite appreciably between the species.

The utilization rate of urea nitrogen by these fungi was not, however, related to the ureolytic activity of the species concerned. Thus though Aspergillus janus has shown the highest rate of urea hydrolysis per hour, it could utilize only a small amount of this nitrogen. Hence, it has been classified as a poor utilizer of urea nitrogen. Penicillium jenseni on the other hand could hydrolyse urea only at a slow rate but this organism successfully utilised about 99% of the nitrogen supplied.

742.
SHENDE, S.T.; APTE, R.G. and SEN, A. (1970) Strain variation in Azotobacter chroococcum in relation to sensitivity to peni-

cillin and mutual antagonism.

Indian J. agric. Sci. 40 : 737-740.

The effectiveness of crop inoculation with Azotobacter is related to the different sensitivity of the strains to toxins produced by other micro-organisms. This was confirmed by the existence of a wide range in sensitivity to penicillin. The strains also showed appreciable antagonism towards one another.

743.
SHENDE, S.T.; BALASUNDARAM, V.R.; BHURAT, M.C. and ABHISWAR SEN (1968) Fixation of atmospheric nitrogen by phyllosphere bacteria. Indian J. agric. Sci. 38 : 298-301.

The amounts of organic acids produced from sugar and the amounts of nitrogen fixed by the bacteria on the leaf surfaces varied in different species of plants. In the case of cauliflower, oat, radish, tomato, barley, mustard and gram, the amounts of nitrogen fixed by the leaf-inhabiting bacteria and of the acid produced were significantly and positively correlated. A significant but negative correlation was observed between the two from bacteria present in the phyllosphere of cauliflower, oat, wheat, potato, lucerne, sugarcane castor, Lantana, Ipomea, pea and berseem. It was possible that different species of nitrogen-fixing bacteria were present in the two groups of plants.

744.
SHINDE, B.N. and ROTE, B.P. (1983) Use of rock phosphate and microbial cultures for the composting of sugarcane trash by heap method. Indian J. agric. Chem. 15 : 147-154.

In an effort to shorten the time required for making good compost, sugarcane trash was composted with different phosphate sources and microbial cultures (Aspergillus sp., Penicillium sp. and Trichurus spiralis) by an open heap method. The results revealed that phosphate sources and cultures combinations markedly hasten the decomposition of sugarcane trash, taking about 3-4 months compared with control which was not ready even after a lapse of 5 months. Composts without P and cultures were not ready even after 5 months.

745.
SHIVAPRAKASHA, M.K. and SHIVAPPA SHETTY, K. (1983). Distribution of N_2-fixing blue-green algae in paddy soils of Shimoga district of Karnataka and their growth rate and N_2-fixing capacity. J. Soil Biol. Ecol. 3 : 77-82.

Twenty seven soil samples from paddy fields of Shimoga were collected and examined for blue-green algae. Out of 21 species isolated, eleven belonged to *Anabaena*, seven to *Nostoc*, and remaining each belonged to *Calothrix*, *Cylindrospermum* and *Scytomema*. It was observed that *Anabaena* variabilis has a maximum growth rate. *Calothrix* sp. fixed maximum amount of nitrogen.

746.
SHRIKHANDE, J.G. (1940) Effect of narrow carbon-nitrogen ratio and of naturally occurring tannins in decomposing plant materials upon the production of mucus. Soil Sci. 49 : 1-8.

Decomposition has been studied from the point of view of production of stickiness, which contributes to good tilth. If environmental pH is suitable, the appearance of stickiness during compost fermentations is not dependent on the nature of the plant material, except that tannin-bearing materials such as tea refuse and *Gordonia* inhibit the production of mucus.

747.
SHRIKHANDE, J.G. (1947) The biological decomposition of green manures. IV - Loss of lignin during aerobic fermentation. Indian J. agric. Sci. 17 : 32-33.

Loss of lignin was small during a period of 5 weeks and during longer periods of fermentation upto 6 months the losses of lignin were only about one-third of those of cellulose and hemicelluloses. In tannin materials loss of lignin was almost negligible.

748.
SHRIKHANDE, J.G. (1949) Effect of naturally occurring tannins on normal process of decomposition. Proc. 4th intern. Cong. Microbial, 1947 p. 481-482.

Tannin materials with over 2% of N have an inhibiting effect on ammonification. Manure, which contains phenolic bodies may exert a similar reaction.

749.
SHUKLA, A.C. and GUPTA, A.B. (1967) Influence of algal growth-promoting substances on growth yield and protein contents of rice plants.

Nature. 213 : 744.

Seeds were pre-soaked for 24 hr in 0.5, 1 and 5 per cent water extract or other extract of Phormidium foveolarum suspended in water, and were sown and raised to maturity. Treatments considerably increased growth and yields and increased protein content of grain. Protein content of grain was increased by the treatments, particularly by the 5 per cent water extract.

750.
SHUKLA, A.N. and DWIVEDI, R.S. (1979) Survival of Rhizoctonia solani Kiihn under the influence of staling growth products of some Aspergilli and its growth response to some phenolic substances.

Proc. Indian Nat. Sci. Acad. 45 B : 269-272.

Rhizoctonia solani causing dampingoff disease in various plants, was isolated from forest soil and its ability to withstand the staling growth products (SGP) of Aspergillus niger, A. flavus and A. candidus was tested on nutrient agar medium amended with the metabolite at a ratio of 1 : 1. Similarly, the capacity of pure culture of the test pathogen to withstand immersion in SGP was studied. All the culture filtrates checked growth of the test pathogen. In another experiment, replacing the carbon source of the medium with phenolic substances (vanillin, ferulic acid and p-hydroxybenxaldehyde) decreased the hyphal dry weight yield and the linear extension of the test fungus.

751.
SHUKLA, U.C. and YADAV, O.P. (1982) Effect of phosphorus and zinc on nodulation and nitrogen fixation in chickpea (*Cicer arietinum* L.).

Trans. 12th int. Congr. Soil Sci. 6 : 54.

A greenhouse study was conducted to study the effect of P and Zn on nodulation and N fixation in chickpea (*Cicer arietinum* L.) grown in a loamy sand (Typic Torripamments). Treatment combinations included five levels of P (0, 25, 50, 100 and 250 ppm) and six levels of Zn (0, 5, 10, 20, 40, 100 ppm). Plants were harvested 64 and 108 days after sowing, and at maturity. Nodule number and dry matter at 64 and 108 days after sowing ranged from 0.5 to 49.5, and 9.6 to 51.0 per plant, and 0.02 to 0.62 and 0.16 to 2.02 g/four plants, respectively. N fixation at 64 and 108 days after sowing, and at maturity ranged from 0.05 to 17.8, 18.7 to 155.8 and 58.6 to 298.2 kg/2 x 10^6 kg soil, respectively. The number, dry matter and leghaemoglobin content of nodules, and N fixed increased with Zn alone upto 10 ppm and P alone upto 50 ppm, and decreased with their higher levels. Maximum nodulation and N fixation (91 to 145 per cent over zero P and Zn, at maturity) was recorded with 25 to 50 ppm P applied along with 5 to 10 ppm Zn. At 64 days a depletion in soil-N was noted, particularly when P was applied whereas, at maturity there was a gain in soil-N regardless of P and Zn treatments. The actual gain depended on the stage of plant growth at harvest, plant residues left in the soil, and N excreted during plant growth. Depending upon P and Zn levels, the gain in soil-N at maturity varied from 10.5 to 44.5 kg/2 x 10^6 kg soil. The increase in nodulation and N fixation with balanced P and Zn nutrition was attributed to increased production of leghaemoglobin and K and Fe concentration in nodules, and increased plant growth, resulting in enhanced activity of N fixing organisms. The results showed that balanced P and Zn nutrition is essential not only for growth but also for maximum activity of Rhizobium for N fixation.

752.
SIDDARAMAPPA, R.; SHANTARAM, M.V. and BALASUBRAMANIAN, A. (1973) The effect of a permanent manurial and cropping schedule on the nitrification process in soil.

Madras agric. J. 60 : 1085-1086.

Maximum nitrification occurred in soils treated with $(NH_4)S_2O_4$ every year, followed by soils treated with farmyard manure and superphosphate.

753.
SIDHU, G.S.; SINGH, N. and SINGH, R. (1967) Symbiotic nitrogen fixation by some summer (kharif) legumes of Punjab. 1. Role of leghaemoglobin in nitrogen fixation.

J. Res. PAU 4 : 244-248.

Total amount of leghaemoglobin contained in nodules of six legumes was directly related to the total amount of N fixed.

754.
SIMHADRI, P. and TILAK, K.V.B.R. (1976) Comparative performance of different strains of Rhizobium species on pigeonpea (Cajanus cajan) (L.) Milsp.).

J. Res. Pantnagar 1 : 26-29.

Pigeonpea seeds were inoculated with nine strains of Rhizobium and grown in pots. Grain yield, nodule number and leghemoglobin content of nodules were all increased by inoculation, but the strains varied very widely in their effectiveness. Effectiveness as well as efficiency in N fixation were greatest for a University of Illinois strain and a peanut strain.

755.
SINGH, A.; RAMAKRISHANA, T. and MURTHY, M.S. (1980) Some ecological observations on two agarophytes from India. Hydro-

biologia, 75 : 185-188.

The responses of Gracilaria corticata and Gelidiopsis gracillis to different concentrations of seawater, pH range and lights of different colours were studied to understand the ecological amplitudes. Both algae had a very narrow range of salinity tolerance from 29.0 to 33.1%, and they could thrive only under pH values from 7.0 to 8.0. Growth was retarded under all coloured lights. However, under yellow light growth was nearer to that of controls.

756.
SINGH, A.P. and BHARGAVA, S.N. (1981) Survival in soil of three species of Fusarium causing wilt of pigeonpea. Phytopath. Z. 100 : 300-311.

Fusarium acuminatus, F. oxysporum and F. solani were commonly found in soils on which pigeonpea (Cajanus cajan) was grown. All three species caused wilt of pigeonpea. Populations of these Fusarium species were highest at 30% water holding capacity of the soil and at soil temperatures of 20-30^{0}C. Addition of carbohydrates to the soil at a 1% rate decreased these populations within 2 and 3 weeks; this was especially marked with sucrose. Application of ammonium sulphate or urea at 0.94 mg N/g (w/w) air-dried soil decreased the populations. Sodium nitrate however increased the population of F. oxysporum.

Populations of all three species decreased with increasing C : N ratio in the soil. Application of Difolatan at 100 ppm decreased these populations.

757.
SINGH, B.D.; MURTY, B.K.; SINGH, R.P. and SINGH, R.B. (1983) Variability and combining ability analysis for symbiotic nitrogen fixation in mung bean (Vigna radiata). Proc. Natl. Symp. Adv. Front. Pl. Sci. Jodhpur. p 259-260.

The efficiency of the Rhizobium - legume symbiosis is significantly affected by genotypes of the Rhizobium and the legume host, and by an interaction among them as well as with the various factors of the environment. This study confirms the suggestion that N-fixation in legumes can be improved by breeding.

758.
SINGH, B.N. (1942) Selection of bacterial food by soil flagellates and amoebae. Ann. Appl. Biol. 29 : 18-22.

Since it appeared that some plant pathogenic bacteria are eaten by protozoa, it is concluded that the incidence of bacterial diseases of plants may be influenced by the presence of protozoa in the soil.

759.
SINGH, B.N. (1945) The selection of bacterial food by soil amoebae, and the toxic effects of bacterial pigments and other products on soil protozoa. Brit. J. Expt. Path. 26 : 316-325.

Amoebae are shown to be very selective as regards the type of bacteria which they eat. Some bacteria are inedible, others are toxic towards protozoa.

760.
SINGH, B.N. and NAIR, K.M. (1939). Is sunlight a factor in nitrogen transformation in soil? Soil Sci. 47 : 285-291.

It is concluded that, in India, nitrite formation is to some extent photochemical, but ammonification and nitrification are mainly biological.

761.
SINGH, B.N.; SINGH, S.N. and NAIR, K.M. (1939) The influence of moisture content and nitrate feeding on the nitrogen cycle in the soil both under aerobic and anaerobic conditions. Proc. Indian Acad. Sci. 9B : 331-342.

A method for measuring ammonification is proposed. There is a particular range of moisture content, both for optimum ammonification and nitrification, being between 34 and 51 per cent of the moisture holding capacity of the soil. Nitrates increased the activity of the ammonifying organisms till a particular concentration is reached. They are more active under aerobic conditions.

762.
SINGH, B.R and TANEJA, S.N. (1977) Effect of gypsum on mineral nitrogen status in alkaline soils.

Pl. Soil. 48 : 315-321.

While monitoring the changes in mineral N status during incubation, it was observed that the magnitude and rate of N mineralization directly depended on organic C and total N contents in the alkaline soils. N mineralization commenced at the beginning of incubation in all except the Hansi and Karnal soils which showed a lag of seven days. These two soils were of high electrical conductivity (EC) and gypsum requirement (GR); N mineralization in them started immediately after amendment with gypsum. It was concluded that the lag was due to the depressive effects of high amounts of toxic salts on microorganisms involved in N mineralization. The gypsum-amended soil samples released more ammonium-N and nitrate-N than unamended samples. Beneficial effects of gypsum were thought to be due to stimulation of microorganisms and/or neutralization of toxic salts.

763.
SINGH, B.S. and SAKSENA, S.B. (1980) Fungistasis due to volatile inhibitors produced by some Penicillium species. Curr. Sci. 49 : 515-517.

Volatile compounds produced by five selected species of Penicillium had fungistatic effects on some pathogenic and saprophytic fungi.

764.
SINGH, C.S. (1970) A study on rhizosphere mycopopulation of some Pteridophytes.

Ann. Inst. Pasteur, Paris. 118 : 825-831.

There was a marked quantitative difference between the rhizosphere and non-rhizosphere flora of 5 Pteridophytes. A qualitative difference was observed only in the case of one (*Dryopteris*) of the plants.

765.
SINGH, D.B.; SINGH, S.P. and GUPTA, R.C. (1979) Antifungal effect of volatiles from seeds of some umbelliferae.

Trans. brit. Mycol. Soc. 73 : 349-350.

In culture medium, volatile substances from germinating seeds of *Anethum graveolens, Cuminum cyminum,* and *Trachyspermum ammi* inhibited all the fungal species isolated from their rhizospheres, whereas the volatile substances from germinating seeds of *Foeniculum vulgare* stimulated radial growth of the fungi. The most inhibitory volatile substances were those from *T. ammi.*

766.
SINGH, G. and SHRIKHANDE, J.G. (1953) Non-symbiotic nitrogen fixation by *Azotobacter.* I - Effect of phosphorus on nitrogen fixation.

J. Indian Soc. Soil Sci. 1 : 47-54.

Experimental evidence has been adduced to show that phosphorus in different forms stimulated nitrogen fixation by *Azotobacter* sp. upto a definite concentration beyond which nitrogen fixation was adversely affected. A difference was observed in the phosphorus requirements of the old and new culture of mixed species of *Azotobacter,* the old requiring higher concentration.

The differences in the optimum requirement of phosphorus from potassium phosphate and superphosphate was due to differences in the availability of phosphorus from the two sources. The manganous phosphate inspite of its limited solubility as compared to that of superphosphate, exerted a stimulating effect on nitrogen fixation upto 2 mgm of phosphorus, beyond which the nitrogen fixation was adversely affected due to increased concentration of manganese which become toxic.

The P/N ratio wideness beyond a optimum concentration of phosphorus showing that the efficiency of Azotobacter is lowered probably because of toxic effect of high concentration of phosphorus and also due to change in hydrogen ion activity.

767.
SINGH, H.N. and SINGH, H.N. (1983) Loss of aerobic nitrogen fixation associated with Kasugamycin resistance in the heterocystous nitrogen-fixing, streptomycin resistant blue-green alga Nostoc muscorum. Plant Cell Reports 2 : 310-313.

Heterocystous N_2-fixing, streptomycin resistant (Het+ Nif+ Str-R) N. muscorum produces two groups of Kasugamycin resistant (Kas-R) phenotypes : one group showing normal heterocyst frequency and normal aerobic growth with N_2 as a nitrogen source, the other group showing loss of growth with N_2 as nitrogen source under aerobic growth conditions.

768.
SINGH, H.P.; PAREEK, R.P. and SINGH, T.A. (1984) Solubilization of rock phosphates by phosphate solubilizers in broth.

Curr. Sci. 53 : 1212-1213.

Among all the P-solubilizers studied, Aspergillus awamorii was the most efficient solubilizer and gave 180 /µg P per 50 ml broth while P. striata was the least efficient giving only 20 /µg per 50 ml broth. Critical examination of the results revealed no correlation between the degree of solubilization and change in pH.

769.
SINGH, I. and PRASAD, S.K. (1978) Influence of pesticides on rhozosphere microbial population of mung and wheat crops.

Indian J. Nematology 8 : 102-109.

Various nematicides were tested for their effects on bacteria, fungi and actinomycetes in the rhizosphere of Phaseolus aureus and wheat. Build up of microbial populations was inhibited to the greatest degree by dazomet and D-D had the least effect. Intermediate effects were observed with DBCP, aldicarb, carbofuran and fensulphothion.

770.
SINGH, J. (1939) A contribution to our knowledge of the Indian soil actinomycetes. Proc. Indian Acad. Sci. B 9 : 69-73.

The actinomycetes do not show any periodic phenomena such as are found in other group of soil organisms.

771.
SINGH, J.N.; TRIPATHI, S.K. and NEGI, P.S. (1974) Yield and nitrogen content of non-nodulating soyabean grown in association with nodulating lines. Proc. Indian natn. Sci. Acad. 40B : 507-511.

Field experiments were conducted for two years at Crop Research Station of the G.B. Pant University of Agriculture and Technology, Pantnagar, to study the associated effect of nodulating soyabean on non-nodulating ones. Isogenic nodulating and non-nodulating soyabean were grown together in separate rows, as well as mixed in the same row in different combinations. Additional treatments of pure nodulating and non-nodulating soyabeans with varying levels of applied nitrogen were also included in the trial. The yield, seed weight and nitrogen content of non-nodulating isogenic genotype increased, when grown in association with nodulating soyabean, over the pure non-nodulating soyabean. The amount of nitrogen obtained from nodulating to non-nodulating in each combination was estimated by using yield and nitrogen data and also by comparing with different levels of applied nitrogen. Even application of 200 kg N/hectare to the non-nodulating soyabeans did not yield at par with nodulating soyabean.

772.
SINGH, J.V. and PUNDARIKAKSHUDU, R. (1984) Studies on

the efficacy of Azotobacter on Nitrogen utilization in rainfed cotton. Golden Jubilee Seminar on Soil Resources and Productivity Management, Indian Society of Soil Science, New Delhi 3 : 57-58.

A field experiment was conducted for three years (1980-82) to study the effect of Azotobacter (C2) inoculation with or withour N on growth, seed cotton yield and nitrogen build-up in soil in rainfed cotton (cv. SRT 1, G. hirsutum). Azotobacter (seed treatment) in combination with N at all the levels (0, 20, 40, 60 & 80 kg/ha) increased the dry matter yields (above ground parts) at maturity over nitrogen alone. Variable response to Azotobacter application with respect to seed cotton yields were observed during different years. However, Azotobacter in combination with nitrogen particularly at lower level i.e. 40 kg/ha improved seed cotton yields over nitrogen alone at the same level which was at par with 60 kg/ha of nitrogen. The available nitrogen content in the soil (0-20 cm) in general, was observed to have been increased with Azotobacter in combination with nitrogen over nitrogen alone upto 45 days after germination. The active process of nitrogen mineralization in Vertisol continued upto 105 days after germination as is evident from the available nitrogen content of the soil beyond which the rate of mineralization seems to have slowed down due to less of soil moisture at the boll development stage. More or less similar trends were observed with respect to NH_4^+ -N and NO_3^- content of soil. Azotobacter application through seed treatment has increased nitrogen uptake in cotton over soil application.

773.
SINGH, K. and SINGH, C.S. (1970) Rhizosphere mycoflora of virus infected grass (Cynodon dactylon Linn).

Acta Biol. 11 A : 359-364.

The percentage occurrence of fungi in the rhizosphere, non-rhizosphere and rhizoplane regions of mosaic-virus infected and healthy plants is shown in tabular form. 19 fungi (2 Phycomycetes, 14 Deuteromycetes, 1 Basidiomycete and, 2 sterile forms) were isolated from the rhizosphere of diseased plants, and 16 fungi (2 Phycomycetes and 14 Deuteromycetes) from the rhizosphere of healthy plants.

774.
SINGH, K.; RAO, K.L.; JAMWAL, J.S. and BOHRA, J.S. (1983) Effect of phosphorus and Bio-fertilizers on growth, yield and quality of chickpea (Cicer arietinum Linn) Trans Isdt & Ucds 8 : 33-38.

Field experiments were conducted during rabi 1978-79 and 1979-80 at the Research Farm Institute of Agricultural Sciences, BHU, Varanasi to evaluate the effect of phosphorus, Rhizobium and phosphobacterin with respect to growth, yield and quality of chickpea. Growth, yield and quality of crop increased with the increasing levels of phosphorus and the highest yield was obtained with the application of 60 kg P_2O_5/ha. Rhizobium inoculation also significantly increased the growth, grain yield and quality of the produce. However, phosphobacterin failed to exert any significant influence on growth, yield and quality of chickpea.

775.
SINGH, H. and MEHTA, V.S. (1964) Study of the effect of chlorinated insecticides on nitrification of alluvial soils of Gwalior. Allahabad Fmr. 38 : 269-271.

On sandy loam soil, pesticide application had no appreciable effect on nitrification. Lindane, aldrin and chlordane had a retarding effect at the beginning, but the effect was not pronounced later.

776.
SINGH, M. and PATHAK, A.N. (1962) Effects of trace elements on nitrogen fixation. Agra Univ. J. Res. 11(3) : 29-36.

Trace elements at 0.1 and 0.2 ppm were added to Azotobacter cultures grown on Ashley's liquid culture medium and soil.

N fixation was determined after 7, 14, and 21 days at 30°C. N fixation was better in the younger culture and better in the soil than in liquid cultures. On the 7th day NO_3-N in the soil started to decrease, especially in the presence of Mo. In both the liquid and soil culture the C : N ratio decreased as decomposition progressed. Mo was most inducive to N fixation. With a 0.2 ppm addition of Mo production of N fixation was highest (0.0204 g N/100 ml medium). Cu, Zn and B additions were not as effective. Co induced the lowest N fixation.

777.
SINGH, N. and SINGH, R.S. (1980) Inhibition of _Fusarium oxysporum_ f. sp. _udum_ by soil bacteria.

Indian Phytopath. 33 : 356-359

In a laboratory experiment _Bacillus subtilis_ and _Bacillus cereus_ inhibited the growth and spore germination and caused lysis of the mycelium and germ tubes of _Fusarium oxysporum_ f. sp. _udum_. _Bacillus badius_ and _Pseudomonas fluorescens_ inhibited spore germination.

778.
SINGH, P. (1954) Symbiotic nitrogen fixation. _Agric. Allahabad_ 4 : 9-12.

The various hypothesis on the biochemistry of symbiotic N fixation are discussed.

779.
SINGH, P.K. (1983) Asymbiotic and symbiotic algal nitrogen fixation - biofertilizer technology for rice. _Proc. Natl. Symp. Adv. Front. Pl. Sci. Jodhpur._ 9 : 105-107.

Nitrogen fixation in paddy field ecosystem by blue-green algae and more recently by _Azolla-Anabaena_ association is estabished and found to be practicable in India because paddy field environment is most suited for their growth and N-fixation. The comparative trials conducted at CRRI farm reveal that rice crop response to _Azolla_ inoculation is better than blue-green algal inoculation.

780.
SINGH, P.N. (1977) Effect of root exudates and extracts of Solanum nigrum and Argemone mexicans seedlings on rhizosphere mycoflora.

Acta Bot. Indica. 5 : 123-127.

Both root exudates and extracts of Solanum nigrum and Argemone mexicana showed varying effects on the growth of fungi in culture. Root extracts of S. nigrum checked the growth of Trichoderma lignorum; root exudate had no marked effect. An alkaloid (solanine) extracted from the roots of the plants was found to be responsible for reduced growth of the fungus. Mucor luteus, Aspergillus flavus, Fusarium solani and F. chlamydosporum showed better growth in root extracts and exudates of S. nigrum compared with the control. Root exudates and extracts of A. mexicana had no significant effect on the fungi studied, except in the case of Cladospium herbarum and Curvularia lunata, which grew better in the presence or root extracts.

781.
SINGH, R. and CHOWDHURY, J.R. (1983) Effect of Rhizobium and common bean mosaic virus infection on the nitrate reductase activity of mung bean.

Sci. Cult. 49 : 366-367.

The results of this study show that Rhizobium treatment enhanced the activity of enzyme more in soil grown plants than those grown in sand. Virus infection further enhanced the NR activity.

782.
SINGH, R.; SINGH, N. and SIDHU, G.S. (1968) Symbiotic nitrogen fixation by some summer (kharif) legumes of Punjab.

J. Res. PAU, 5 : 88-94.

In pot experiments with Sesbania aculeata, S. macrocarpa, S. sericae, Crotalaria juncea, Vigna sinensis and Cyamopsis tetragonolobus, super increased the yield of tops, roots, nodules, and the total N content of the plant tissue. Sesbania species gave the highest yield and total N content. The other species gave greater response to super than Sesbania did.

783.
SINGH, R.K.; SINGH, R.B.; SINGH, R.M. and SINGH, B.D. (1980) Nitrogen fixation by nitrate reductase deficient mutants of Rhizobium japonicum.

Indian J. exp. Biol. 18 : 1165-1167.

Forty mutants resistant to chlorate (10 mg/ml) were isolated from Rhizobium japonicum CB 1809; the mutants were unable to reduce nitrate used as sole source of nitrogen. Based on the induction of nitrate reductase activity in nitrate and nitrate + glutamate containing minimal medium, these mutants were categorized into 2 classes. Mutants of Class I showed a marked increase in nitrate reductase activity in the glutamate supplemented nitrate medium in comparison with the medium containing nitrate only. Mutants of Class II were unable to reduce nitrate in either medium. Apparently Class I type mutants were defective in the regulation of nitrate reductase, while the Class II type were probably affected at the structural genes, nar. Five mutants of each class were tested for nodulation and nitrogen fixation on soyabean plants; they were infective and their efficiency was comparable to CB 1809. Thus nitrate reductase deficient mutants do not have impaired nitrogenase activity. It is concluded that the 2 enzymes do not share a common genetic determinant.

784.
SINGH, R.N. (1942) The fixation of elementary nitrogen by some of the commonest blue-green algae from the paddy field soils of the United Provinces and Bihar. Indian J. agric. Sci. 12 : 743-756.

Laboratory study suggest that nitrogen recuperation in these soils is an algal process in which several spices take part. A grass-green alga fixed no nitrogen. The N-fixing algae are unique in being able to obtain both their C and N from the air; this accounts for their ability to colonize new soils.

785.
SINGH, R.N. (1950) Reclamation of usar lands in India through blue-green algae. Nature, 165 : 325-326.

The species of blue-green algae found in alkaline usar soils of northern India are listed. Through *Nostoc commune* "the annual addition of organic matter of the soil is of the order of 36.5-59.7 per cent and that of nitrogen 30-38.4 per cent" and N fixation by it is strongly suggested by preliminary investigations. *Azotobacter* populations are reduced. Reclamation is confined to the rainy season but could be extended with irrigation. Earth embankments are made to enclose plots and algae develop in succession as the water table rises. Paddy and sugarcane crops have been successfully grown on second-and-third year reclaimed land. The soil improves in pH, tilth and exchangeable-Ca content and also markedly in water-holding capacity.

786.
SINGH, R.N. (1978) The biology of nitrogen fixation by blue-green algae. pp. 288-289. In S.P. Sen, Y.P. Abrol and S.R. Sinha (eds.), Nitrogen Assimilation and Crop Productivity. Associated Publishing Company, New Delhi.

In view of the occurrence of a large number of nitrogen-fixing and photosynthetic blue-green algal endosymbionts in the tissues of higher plants, and the revival of the theory of origin of higher plant chloroplasts from some such endosymbionts, it may be possible to transfer a block of genetic information from blue-green algae capable of photosynthetic nitrogen fixation, to chloroplast of cereal crops like rice in haploid tissue cultures either by transformation, transduction (through cyanophages) or by fusion of isolated protoplasts. Biological nitrogen fixation presently has entered a very exciting phase of genetic analysis. The purpose of this paper is to highlight the genetic regulation of the nitrogen-fixing, *nif* gene in blue-green algae--the key organisms in the overall economy of nitrogen in crop production.

787.
SINGH, R.P.; LOKENDRA SINGH and VERMA, V.S. (1984) Effect of *Azotobacter* exudates on certain micro-fungi associated with the phylloplane of soyabean.

Phytopath. Z. 109 : 97-103.

It was found that the growth of some fungi was stimulated or suppressed, whereas others showed fluctuations or remained unaffected depending upon the age of the plant, growth stage of the leaf, concentration of *Azotobacter* exudates, method used to harvest the fungi and environmental conditions. Nutrient competition or production of biologically active substances like GA may be responsible for retarding the growth of fungi.

788.
SINGH, R.S. (1984) Significance of nitrification inhibitors, slow release N-fertilizers and biofertilizers in nitrogen management in rice field.

Farmer & Parliament 19(10) : 13-14.

The use of nitrification inhibitors, slow release fertilizers and biofertilizers minimize the loss of nitrogen, and consequently enhance its use efficiency in rice fields considerably where nitrogen losses occur heavily. Accordingly they have significant positive effect on yield of rice grain.

789.
SINGH, R.S. and KHANNA, R.N. (1977) Population dynamics of *Fusarium* sp. in pea rhizosphere as affected by soil amendments.

Folia Microbiol. 22 : 396-401.

In nonamended soil, vegetative growth and sporulation of *Fusarium* spp. were higher in the rhizosphere than in the soil. Sporulation was favoured by young plants and decreased with increasing plant age. Amendments with low C/N oilcakes enhanced vegetative growth and sporulation in root-free soil. The extent of stimulation varied with the nature of the organic matter used and the stage of its decomposition. Sporulation was suppressed by castor cake, and by sawdust with urea amendments. The rhizosphere altered the effect of the different amendments.

790.
SINGH, R.S. and REDDY, C.S. (1979) Suppression of damping-off of tomato and seedling blight of chick pea and sugarbeet by strains of Streptomyces diastaticus.

Indian Phytopath. 32 : 374-377.

An isolate of Streptomyces diastaticus suppressed the growth of Pythium aphanidermatum in autoclaved as well as natural soil in presence of tomato seeds and seedlings by about 50 per cent within 20 days. The incidence of post-emergence damping-off was also reduced by more than 50 per cent but pre-emergence damping off was not affected as much. In the subsequent planting, immediately following the first crop, the antagonist caused significant reduction of both pre-as well as post-emergence damping-off. Another isolate of S. diastaticus reduced seedling blight of chick pea (Sclerotium rolfsii) by 86 per cent in autoclaved soil and by 59 per cent in natural soil. Seedling blight of sugarbeet (S. rolfsii) grown in the same soil immediately after chick pea was reduced by 83 and 70 per cent in autoclaved and natural soil, respectively.

791
SINGH, R.S. and SINGH, R.P. (1984) Response of Pea (Pisum sativum) to sulphur fertilization and Rhizobium inoculation in different soils of Uttar Pradesh. Golden Jubilee Seminar on Soil Resources and Productivity Management, Indian Society of Soil Science, New Delhi 4 : 84.

Grain and straw yields significantly increased due to sulphur fertilization and Rhizobium inoculation. Highest grain and straw yields were obtained at 30 ppm S in both the years. Higher yields were recorded in red soil followed by alluvial, calcareous and mar (clayey) soils. The magnitude of response to inoculation on grain and straw yields was reckoned to be 8.45 and 10.76 per cent and 6.32 and 11.26 per cent in the crop seasons of 1981-82 and 1982-83, respectively.

The contents of N, P, K, Ca and Mg were in general, not affected by different treatments viz., sulphur application, soil types and Rhizobium inoculation, whereas the sulphur content was significantly influenced by its application. Uptake of all the nutrients was significantly enhanced as in case of yield with these treatments.

792.
SINGH, S. and RAI, R.N. (1975) Effect of salinity, alkalinity, phosphate and age of plants on mineralisation of nitrogen from Sesbania aculeata and Melilotus alba after their incorporation in soil. J. Indian Soc. Soil Sci., 23 : 122.

Salinity and alkalinity of soils depressed the mineralisation of nitrogen during decomposition of both the plants. Nitrogen losses were higher under alkaline condition of the soil. Application of phosphate enhanced mineralisation but reduced nitrogen losses. Mineralisation percentage was higher in the soils in which 60 days old plants were allowed to decompose. In the soils having 75 days old Sesbania aculeata and 90 days old Melilotus alba, ammonium N was lost after 30 days of incubation, while in the soils with young plants it was after 15 days. Total nitrogen was more reduced after 15 days of incubation in soils with young plants.

793.
SINGH, S.B. and PATHAK, A.N. (1962) Effect of trace element on organic matter decomposition. Agra Univ. J. Res. 11 : 97-103.

In laboratory experiments normal cultivated soil was treated with sanai at 150 maunds/acre and Mo, B, Zn, Cu or Co was added at 0.1 or 0.2 ppm. Organic matter decomposition, N fixation, available P and NO_3 contents and aggregation were increased by trace-element treatment. Mo had the strongest effect.

794.
SINGH, S.J. and TEWARI, R.P. (1979) Rhizosphere mycoflora of healthy and yellow vein mosaic virus infected okra (*Abelmoschus esculentus*) plants.

Acta Microbiol. Polonica 28 : 311-314.

Investigations on the rhizosphere mycoflora of healthy okra plants and okra plants infected with yellow vein mosaic virus showed a higher fungal population in the rhizosphere of healthy plants at preflowering and postflowering stages than in that of diseased ones. Maximum populations were observed during flowering in all cases, including non-rhizosphere soil, but in virus infected plants at this stage the rhizosphere populations were higher than in healthy plants, apparently due to differences in C/N ratio and amino acids. A drastic reduction in the diseased plant rhizosphere occurred during the post-flowering stage however, which may be due to an unfavourable change in C/N ratio, or to the production of some toxic substances inhibiting multiplication of the mycoflora.

795.
SINGH, S.N. (1982) Effect of foliar spray on colonization of microflora under the rhizosphere of crop plants.

J. Indian Bot. Soc. 61 : 323-325.

When *Phaseolus aureus (Vigna radiata)* plants were sprayed with IAA at 15-day intervals, fungal and bacterial populations in the rhizosphere increased with concentrations up to 100 ppm but decreased with higher concentrations.

796.
SINGH, S.N. and SARBHOY, A.K. (1976) *Syncephalis indica* sp. n. from India.

Zentbl. Bakt. ParasitKde. Abt. II. 131 : 201-204.

A description is given of this fungus which was isolated from heavily manured soil at Jodhpur, Rajasthan as a hyperparasite on Mucor sp.

797.
SINGH, T.H. and IYENGER, M.R.S. (1960) Nitrification by soil fungi. Arch. Mikrobiol. 35 : 252-257.

The nitrifying fungi, isolated by the dilution plate method on Czapek-solution agar or potato-dextrose agar, belonged to the genera Aspergillus and Penicillium, isolates of the former producing both NO_2 and NO_3, and those of the latter producing only NO_3 from NO_2.

798.
SINGH, U.R. (1977) Relationship between the population density of soil microarthropods and mycoflora associated with litter and the total litter respiration on the floor of a sal forest in Varanasi, India.

Ecol. Bull. (Stockholm) 25 : 463-470.

Litter respiration under Shorea robusta, as measured by the alkali absorbtion method, was highly correlated with (1) the microarthropod population ($r = 0.903$; $P < 0.01$) and (2) the mycofloral population ($r = 0.79$; $P < 0.1$). An allometric equation was developed for both the parameters i.e. mycoflora versus total litter respiration ($y = 0.4425 + 0.601\ x$) and microarthropods versus total litter respiration ($y = 12792 + 18\ x$) in order to estimate one variable from a knowledge of the other. Ecological factors, such as rainfall, humidity, temperature (air & litter), litter moisture, pH (litter) were also recorded, and had significant effects on the biological activity of the soil.

799.
SINGH, U.R. and SHUKLA, A.N. (1977) Soil respiration in relation to mesofaunal and mycofloral populations during rapid course of decomposition on the floor of a tropical dry deciduous forest.

Rev. Ecol. Biol. Sol. 14 : 363-370.

CO_2 evolution from the floor of a tropical dry deciduous sal forest was measured *in situ*. Significant correlations were found between CO_2 evolution and (a) numbers of mesofauna (Tullgren funnel method) and (b) mycofloral populations (plate counts).

800.
SINGH, V.P. (1978) Mutations defective in nitrogen fixation, heterocyst and spore production in the blue-green alga *Anabaeua variabilis*.

Biochem. Physiol. Pflanzen. 173 : 365-372.

N-methyl-N'-nitro-N-nitrooguanidine (NTG)-induced mutants of the filamentous, heterocystous, nitrogen fixing blue-green alga *Anabaena variabilis* are classified into five groups : (i) Non-nitrogen fixing with normal trichome, (ii) non-sporulating and colony forming; (iii) non-nitrogen fixing with short trichome; (iv) non-sporulating and non-colony forming and (v) non-sporulating with gas vacuoles. All the mutants were stable and did not revert during successive subcultures. The induction of *nif* mutants showed the maximum mutation frequency amongst all the mutants concerned. The salient features of the mutants are described.

801.
SINGH, V.P. (1979) A novel type of UV action in *Anabaena variabilis* Kiitz.

Biochem. Physiol. Pflanzen 174 : 76-79.

A. variabilis forms distinct pinhead-sized heterocystous colonies on N-free agar medium. UV treatment resulted in the production of a colony showing a spreading tendency and lacking heterocysts. Such colonies grew in N-free medium for 25 to 30 days; thereafter all changed to heterocystous type.

802.
SINHA, A.P.; AGNIHOTRI, V.P. and SINGH, K. (1979) Effect of herbicides on nitrification and ammonification activity in soil.

Indian J. Microbiol. 19 : 52-55.

Alachlor (1.5, 7.5, 15 and 30 ppm) and diuron (1, 5, 10 and 20 ppm) were applied to sandy loam soil to examine their effects on the nitrification and ammonification activities of soil micro-organisms. Recommended rates (2 to 3 kg/ha) of the 2 herbicides did not produce any detrimental effect on the nitrification process throughout the experiment. However, the nitrification process was inhibited by alachlor (15 and 30 ppm) and diuron (20 ppm) for 7 and 30 days, respectively. Normal rates impaired ammonification but higher concns of alachlor (7.5, 15 and 30 ppm) and diuron (5, 10 and 20 ppm) stimulation the ammonification activity for 45 days and the normal rate of diuron slightly stimulated nitrification.

803.
SINHA, A.P.; AGNIHOTRI, V.P. and KISHAN SINGH (1980) Effect of insecticides on the growth and sporulation of rhizosphere fungi of sugarbeet seedlings. Indian Phytopath. 33 : 96-98.

Of the three insecticides evaluated, lindane was more toxic than the organophosphate insecticides dasanit or thimet in inhibiting growth of rhizosphere fungi of sugar beet. Dasanit was much more effective than thimet.

804.
SINHA, A.P.; AGNIHOTRI, V.P. and KISHAN SINGH (1980) Influence of the herbicide alachlor on soil microorganisms and carbon dioxide production. Proc. Indian natn. Sci. Acad. 46 B : 562-565.

At the customary rate of usage (1.5 ppm), alachlor had no detrimental effects on soil fungi, actinomycetes and bacteria. The herbicide stimulated soil respiration for the first 20 days.

805.
SINHA, P.; JHA, K.K. and SINGH, R. (1958) Nature of N mineralization and CO_2 evolution in some typical Bihar soils. Proc. Bihar Acad. agric. Sci. 5 : 32-46.

With 10 upland soils incubated under aerobic conditions at

30°C and 1/3 moisture saturation, N-mineralization was slow for 6 weeks and then rose sharply. CO_2 evolution correlated significantly with N mineralization and organic-C content. No significant correlation with pH or C/N ratio was obtained.

806.
SINHA, P.; JHA, K.K. and SINGH, R. (1959) Nature of nitrogen mineralization and carbon dioxide evolution in some typical Bihar soils.

J. Soil Wat. Conserv. India 7 : 88-97.

N mineralization was slow for the first 6 weeks of incubation and then rose sharply. CO_2 evolution was significantly correlated with N mineralization and with organic matter content, suggesting a similarity in the nature of organic matter in these soils; pH and C/N ratio were not correlated with CO_2 evolution.

807.
SINHA, S.B. (1958) Contribution of *Azotobacter* in nitrogen fixation under natural conditions.

J. Indian Soc. Soil Sci. 6 : 153-159.

The efficiency of *Azotobacter* as nitrogen fixer, has been studied under different conditions and it has been established that the efficiency of *Azotobacter* dies out with time, also that the bacteria are killed when exposed to sunlight. Fixation of nitrogen even then continues in flasks exposed to sunlight.

It has been clearly demonstrated that under tropical conditions, where sunlight is abundant the *Azotobacter* does not play a big role in increasing the nitrogen status of soils.

808.
SINHA, S.B. (1968) Study of nitrification in cultures of nitrifying bacteria.

J. Indian Soc. Soil Sci. 16 : 337-340.

Contrary to the generally accepted idea that bacterial conversion of ammonia to nitrate is practically quantitative, it has been observed that an appreciable amount of nitrogen is lost in the nitrification of ammonium sulphate through bacterial agency just like photo-nitrification or the chemical conversion of ammonia to nitrate. It is postulated that during the process of nitrification, the unstable compound (ammonium nitrite) is produced as an intermediate product which decomposes into water and nitrogen. The loss may also be partially due to the decomposition of nitrous acid, and especially when the pH is less than 7.

809.
SINHA, S.B. and SHARMA, H.G. (1961) Mineralisation of nitrogen in organic manures in Gwalior soil.

J. Indian Soc. Soil Sci. 9 : 201-204.

Among the different organic materials, farmyard manure was found to have a low rate of ammonification and nitrification. In the case of green manure, mineralisation of nitrogen was not found to proceed at a fixed rate, atleast during the period of twelve weeks, but amounts of ammoniacal and nitrate nitrogen after six weeks were quite considerable. Compost showed a higher rate of mineralisation of its nitrogen than farmyard manure. In the case of groundnut cake, the rate of nitrate formation was highest though the formation of ammoniacal nitrogen or mineralisation in general was not found to occur at a fixed rate. For a rapid supply of available nitrogen to the soil, groundnut cake was observed to be the most suitable manure next in order was either compost or green manure, while farmyard manure was least suitable.

810.
SINHA, S.K. (1978) Nodulation, growth and yield in pulse crops. pp.27-43. In S.P. Sen, Y.P. Abrol and S.K. Sinha (eds.), Nitrogen Assimilation and Crop Productivity. Associated Publishing Company, New Delhi.

Investigations on cowpea, mungbean and Bengal gram have

shown that the nodulation keeps pace with plant growth until flowering starts. The photosynthesis rates in these crops are higher in the seedlings and pre-flowering stages. With early pod development, both the nitrogen content of leaves and the rate of photosynthesis fall. Pod development pattern in Bengal gram (Cicer arientinum) showed that the basal pods contained higher grain weight and higher protein content. These components decreased base upward, apparently suggesting limitation of both photosynthates and nitrogen. In most pulses intraplant competition seems to be responsible for poor realisation of 'sink' potential.

811.
SIVAPRASAD, P. and RAI, P.V. (1984) Allantoin content of Cajanus cajan (L.) Mill sp. inoculated with VA-mycorrhiza and rhizobium.

Curr. Sci. 53 : 1249-1251.

Higher allantoin content noticed in mycorrhiza alone inoculated plant over Rhizobium alone or no inoculation show the role of VA-mycorrhiza in bringing about effective symbiosis by native rhizobia; probably due to better nutrient availability conferred by mycorrhizal association.

812.
SOHOO, M.S.; CHAHAL, V.P.S. and BERI, S.M. (1984) Effect of Rhizobium inoculation on different varieties of cowpea (Vigna unguiculata L. Walp.)

J. Res. PAU 21 : 316-319.

Different varieties were found to behave differently to inoculation with the same strain of Rhizobium. Out of eight varieties tested, only one i.e. cowpea 85, formed the best combination with the efficient strain C-41 of Rhizobium used for seed inoculation.

813.
SOMANI, L.L. and SAXENA, S.N. (1970) Acid gradient elution studies on release of phosphorus under active microbiological

activity in Rajasthan soils.

J. Indian Soc. Soil Sci. 18 : 379-382.

Acid gradient elution of fifteen surface soils treated with culture alone and culture alongwith glucose after two months of incubation at 35°C revealed increased release of phosphorus. Glucose additions released more phosphorus. Soils rich in total P, organic P, and having low organic carbon/organic phosphorus ration provided better release.

814.
SOMANI, L.L. and SAXENA, S.N. (1971) Studies on the mineralization of organic phosphorus under the influence of crop growth in some soils of Rajasthan.

J. Indian Soc. Soil Sci. 19 : 261-267.

Studies on the organic C/organic P ratios of some soils of Rajasthan show that they are rather low indicating a high mineralizing capability. It is revealed from pot experiments that Sorghum vulgare and Phaseolus mungo have some kind of dependence on total and hydrolyzable organic P. Mineralization study of rhizosphere and non-rhizosphere soils has indicated that hydrolyzable organic P is not so readily utilized. It is inferred that phytase activity of plant roots helps in an overall release and utilization of organic P.

815.
SOMANI, L.L. and SAXENA, S.N. (1981) Effects of organic and inorganic amendments on the microflora and crop growth in calcareous saline alkali soil. Pedobiologia, 21 : 192-201.

A field trial was laid out to study the role of gypsum and sulphur alone and in conjuction with organic materials, namely, dhaincha, farmyard manure, poultry manure and rice husk in improving the soil microlora and crop growth on a calcerous saline alkali soil. The counts of total bacteria, nitrifiers, Azotobacter sp., clostridia and fungi all increased in direct proportion to the extent of reclamation brought about by the various treatments. The relative proportions of bacteria over actinomycetes, of nitrate former over nitrite former, and of Azotobacter sp. over clostridia greatly increased with the improvements in salt balance and soil physical properties during reclamation. The bacterial spores decreased with reclamation. The improvement in soil microflora was related to the improvement in soil physical and chemical properties and to the growth and yield of wheat. There was greater improvement with sulphur alone or with organic materials than with gypsum. The best treatment was dhaincha + sulphur. Of the organic materials studied, dhaincha gave best improvement in soil microflora and wheat yield. Organic materials and inorganic amendments showed an additive effect in improving soil microflora and crop growth.

816.
SRINIVASAN, K.V. (1968) Effect of a varietal mixture on sugarcane root rot susceptibility and on the rhizosphere microflora.

Proc. Indian Acad. Sci. 58 B : 150-162.

Mixed cultivation of two varieties of sugarcane differing in resistance to root rot intensified the total rhizosphere effect of each of the components and conferred upon the susceptible variety a degree of resistance, apparently by changing the character of the fungal flora. That it was not due to diffusion of antibiotic substances of actinomycete origin from the rhizosphere of the resistant variety was indicated by the results of experiments in which actinomycetes and fungi were separately restored to sterilized soil.

817.
SRIVASTAVA, J.S.; SINGH, B.D.; DHAR, B.; SINGH, V.P.; SINGH, R.B. and SINGH, R.M. (1980) Use of streptomycin resistance and phage sensitivity as markers in competition studies with Rhizobium leguminosarum.

Indian J. exp. Biol. 18 : 1171-1173.

The frequency of nodulation by inoculants was studied in pea using R. leguminosarum strain P5, sensitive to phage RL1, and 4 mutants of P5 resistant to 100/@g/ml streptomycin. The str mutants were comparable or superior to the parent strain in effectiveness and competitive ability in test tube experiments. In the field, the frequency of nodulation by Str mutants ranged from 4.3-12.8%. The low frequency of nodulation by the inoculants emphasizes the necessity for developing Rhizobium strains with superior competitive ability against the native rhizobia. A high proportion (@ 90%) of the streptomycin resistant clones were sensitive to phage RL1 in plots inoculated with str mutants, but none was sensitive in the control or P5-inoculated plots, indicating that phage sensitivity would be a useful marker in ecological studies, either alone or in combination with drug resistance.

818.
SRIVASTAVA, K.S. and SINGH, S. (1978) Citric acid production by three members of Aspergillus ochraceus group.

J. Indian Bot. Soc. 57 (Suppl.) : 7.

Aspergillus lanosus, A. melleus and A. ochraceus isolated from soils of the teak forest of Gorakhpur showed an ability to convert sugar into citric acid at low efficiency. The occurrence of this process of fermentation in members of the Aspergillus ochraceus group has up to now been uncertain.

819.
SRIVASTAVA, L.L. and AHMED, N. (1984) Effect of decomposed straw mulch material on physicochemical and microbiological properties of soil and on crop yield. Golden Jubilee Seminar on Soil Resources and Productivity Management, Indian Society of Soil Science, New Delhi 3 : 55.

In a microplot experiment, use of rice straw as surface mulch at the rate of 100 q/ha spread after sowing of wheat and pea in an acidic red loam soil of Ranchi was found to conserve soil moisture better than in unmulched plots at various stages of sampling. The soil temperature of mulched plots was lower by about 2°C as compared to bare ones. The congenial conditions of soil moisture and temperature created by mulching appeared to be reflected in enhanced microbial population and significantly higher yield of grain and straw of wheat and pea as compared to that in unmulched plots. The straw mulch as a result of partial decomposition was found to slightly increase the total carbon and nitrogen status of soil samples (0-8 cm) collected after harvesting of the crops. The straw mulch turned under the soil after harvest decomposed and enriched the soil which was reflected in significantly higher grain yield of maize sown in the following kharif (monsoon) season.

820.
SRIVASTAVA, O.P.; MUNDRA, M.C.; SETHI, B.C. and KHANNA, S.S. (1974) Contribution of legumes in nitrogen build-up of soil and wheat production. Proc. Indian natn. Sci. Acad. 40B: 516-525.

Jowar and bajra are exhaustive crops with the result that the yield of succeeding wheat is very poor. Green manuring with gaur gives benefit equivalent to 60 kg N application/ha. The yield after guar (for seed) is almost equal to yield after fallow which shows that guar (for seed) is self-sufficient for its notrogen requirement. Fallowing has been observed to decrease the nitrogen content of soil especially the amino-acid-N-fraction. Additions of higher doses of nitrogen have increased the total nitrogen content of soil but not the distribution pattern of various forms of organic nitrogen.

821.
SRIVASTAVA, O.P.; NASIM, S. and KUMAR, J. (1978) A survey of some actinomycetes from tarai soils of U.P. for antifungal property.

J. Indian Bot. Soc. 57 : 24-27.

One hundred and eighteen actinomycetes from tarai soils of Uttar Pradesh were surveyed for antifungal property in the culture filtrates against four ringworm fungi. Trichophyton rubrum, T. mentagrophytes, Microsporum canis and Epidermophyton floccosum. Filtrates from five actinomycete cultures were found to exert significant inhibitory effect against all the four ringworm fungi.

822.
SRIVASTAVA, P.N. and DWIVEDI, V. (1983) Chlorococcales from the soils of Western Rajasthan. Trans. Isdt & Ucds 8 : 93-96.

Eighteen members, belonging to five genera, of chlorococales which developed in the cultures obtained from soil samples, have been described. Characiosiphon rivularis Iyengar grows after the rains and is common at certain places.

823.
SRIVASTAVA, S.C. (1966) The effect of Telodrin on nitrification of ammonia in soil and its implication on nitrogen nutrition of sugarcane.

Pl. Soil 25 : 471-473.

Soil dressed with $(NH_4)_2SO_4$ and treated with the chlorinated hydrocarbon insecticide contained higher amounts of total inorganic N and NH_4-N than untreated plots due to impaired nitrification. This and possibly the effects of other chlorinated insecticides benefit the sugarcane crop and result in increased yields.

824.
SRIVASTAVA, V.B. (1971) Investigations into rhizosphere microflora. VIII. Light and dark treatments in relation to root region microflora.

Pl. Soil 35 : 463-470.

Higher microbial populations occurred in the rhizospheres of wheat and barley plants exposed in continuous light than in those of plants kept continuously in the dark; a 12 hr alternate light and dark treatment had intermediate effects. Amino-acid contents were always higher in the roots of plants exposed to continuous light. Changes in microbial activity were associated with variations in root excretions and the physiological and morphological condition of the plants.

825.
SRIVASTAVA, V.B. and MISHRA, R.R. (1971) Investigations into rhizosphere microflora. I. Succession of microflora on root regions of oryza sativa L.

Microbiologia Espanola 24 : 193-205.

The rhizosphere and rhizoplane microflora of Oryza sativa was studied from the seedling to the decomposing stages. The microflora was more abundant in the crown of the rhizosphere and least in the rhizoplane. Tables show the distribution of fungal species in the different root regions. Curvularia lunata, Alternaria tenuis, Fusarium avenaceum, F. oxysporum and black sterile colonies were dominant in the rhizosphere region.

826.
SRIVASTAVA, V.B. and MISHRA, R.R. (1971) Investigation into rhizosphere microflora. III. Influence of sampling time.

Microbiologia Espanola, 24 : 2.

Sampling period (before, after and during flowering) had little effect on the qualitative composition of the rhizosphere microflora of 2 medicinal plants. Quantitative composition was directly related to the amount of total free amino acids in the roots.

827.
SUBBA RANGAIAH, G.; UMAMAHESWARA RAO, M. and RAO, B.G.S. (1975) Effect of salinity on spore shedding in Gracilaria corticata. Curr. Sci., 44 : 717-718.

Greater numbers of spores were shed at salinities between 30% and 50%, which appeared to be the optimum range for enchanced tetra spore liberation in G. corticata although the peak values were obtained at 40%. Salinities lower than 30% or over 50% were neither conducive nor positively inhibit spore shedding rates under particular temperature and light.

828.
SUBBA RAO, N.S. (1977) Soil Microorganisms and plant Growth. Oxford & IBH Publishing Co. New Delhi, 289 p.

An illustrated book dealing with various types of interactions between soil microorganisms and higher plants and the importance of some microbial metabolites which are relevant to the growth and well-being of plants. Also outlined are certain microbial activities in soil involved in nutrient absorption by plants.

829.
SUBBA RAO, N.S. (1979) (ed.). Recent Advances in Biological Nitrogen Fixation. Oxford & IBH Publishing Co. New Delhi, 486 p.

An illustrated volume reviewing the current status of fundamental and applied knowledge on the subject and also indicates future possibilities with 16 chapters written by experts in their chosen areas. The bibliography is extensive and comprehensive.

830.
SUBBA RAO, N.S. (1982) (ed.). Advances in Agricultural Microbiology. Oxford & IBH Publishing Co. New Delhi, p 704.

This book has 25 chapters contributed by experts in the field from all over the world and has been arranged in three sections: (1) the role of microorganisms in the mobilization of nitrogen and phosphorus to plant growth; (2) management of pathogens, pests and weeds through microorganisms; and (3) new strategies in bioconversion. The chapters cover all areas of soil-plant-microorganisms relationship which in turn are interlinked with the interchange of elements in the biosphere. Some of the chapters emphasize the role of microorganisms in nutrition, pest control and energy recycling.

831.
SUBBA RAO, N.S. (1982) Biofertilizers. In Advances in Agricultural Microbiology. (Ed. by N.S. Subba Rao). Oxford & IBH Publishing Co. New Delhi, p 219-242.

The article provides a comprehensive review on all nutrient inputs for plant growth which are of biological origin viz. Rhizobia, Azotobacter, Azospirillum, Azolla etc. used as microbial inoculants.

832.
SUBBA RAO, N.S. (1982) Biofertilizers in Agriculture. Oxford & IBH Publishing Co. New Delhi, p. 186.

Presented in this well-illustrated book is an up-to-date account of fundamental as well as applied aspects of research on biofertilizers which are finding large-scale application as supplements to chemical fertilizers in the modern-day crop cultivation practices. Biofertilizers can be more appropriately called 'microbial inoculants' and it is now accepted that microbial processes are not only quicker but also consume relatively less energy than industrial processes. Secondly, these processes have the advantage of being diversified into small units to meet the demands of location specific problems related to the agricultural practices of nations which have not yet adopted mechanisation of farming.

The book provides brief, basis and relevant information on the subject with appropriate references for the benefit of both biologist and farm scientists who are interested in alternative sources of energy in agriculture through beneficial microorganisms more particularly in the developing countries.

833.
SUBBA RAO, N.S. (1982) Phosphate solubilization by soil microorganisms. In Advances in Agricultural Microbiology (Ed. by N.S. Subba Rao). Oxford & IBH Publishing Co. New Delhi, p. 295-303.

The article provides a comprehensive account on the role

of phosphate solubilizing microorganisms through acid secretion, the mineralization of organic phosphatic compounds by which organic phosphorus is converted to inorganic ions. Benefit due to inoculation with phosphate dissolving microorganisms have also been given.

834.
SUBBA RAO, N.S. (1982) Utilization of farm wastes and residues in Agriculture. In Advances in Agricultural Microbiology (Ed. by N.S. Subba Rao). Oxford & IBH Publishing Co. New Delhi., p. 509-521.

The article stresses the importance of recycling organic wastes not only to conserve energy but also to minimize pollution. There is an urgent need for the developing countries to accurately assess the availability of farm wastes and the economics of recycling of such wastes both on the small as well as big farms so as to obtain a realistic picture of the type of wastes which remain abundant on the farm for inexpensive recycling operations. The two well-recognized methods of utilizing farm wastes are composting and biogas production. Realistically speaking, organised scientific composting is yet to be taken up on a large scale in developing countries whereas the biogas technology is catching up increasingly due to the massive efforts to popularize biogas plants.

835.
SUBBA RAO, N.S. and BAJPAI, P.D. (1965) Fungi on the surface of legume root nodules and phosphate solubilization.

Experientia 21 : 386-387.

Cephalosporium, Alternaria, Aspergillus and Penicillium occurring on root nodules are capable of solubilizing P. Cephalosporium inhibited growth of two species of P-dissolving bacteria, but alone it released as much P as the individual bacteria did. Penicillium released nearly four times as much P as the individual bacteria did.

836.
SUBBA RAO, N.S. and SARMA, K.S.B. (1968) Pectin methyl esterase activity of root exudates of legumes in relation to rhizobia.

Pl. Soil 28 : 407-412.

Seven different legumes were grown aseptically and in the presence of their specific and non-specific rhizobia. Pectin methyl esterase (PME) was present in all the root exudates from aseptically grown plants but in smaller quantities than were found in plants inoculated with their specific rhizobia. The PME content of exudates from plants inoculated with non-specific rhizobia was always lower than in exudates from plants grown with their specific rhizobia and in some cases was lower than the PME content of uninoculated ones, except in the case of Trigonella foenum-graecum where three non-specific rhizobia gave greater enzyme activity than the specific rhizobia.

837.
SUBBA RAO, N.S. and VASANTHA, P. (1965) Nodulation of Trifolium alexandrinum in vitro and nitrate effect on the amino acid composition of the plant and its root exudate.

Can. J. Bot. 43 : 1189-1194.

A method to bring about root nodulation in vitro by growing berseem plants aseptically and inoculating them with Rhizobium was used to verify the inhibitory effects of excess of combined N in soil on nodulation. Though asparagine was present in plant tissue in relatively higher concentration both in the NO_3- added and NO_3- free series, it was exuded only by roots of NO_3- added series. The presence of tryptophane in root exudates, and its implication in the infection process of legumes by Rhizobium is stressed, since the amino acid is a possible percursor of indolacetic acid which causes curling and deformation of root hairs as a precondition of legume-Rhizobium symbiosis.

838.
SUBBA RAO, N.S.; LAKSHMI KUMARI, M.; SINGH, C.S. and MAGU, S.P. (1972) Nodulation of lucerne (Medicago sativa L.) under the influence of sodium chloride. Indian J. agric. Sci. 42 : 384-386.

Strains of Rhizobium meliloti nodulating the plants of lucerne tolerated sodium chloride up to 3.0 per cent. However, seeds of lucerne did not germinate even at 1.5 per cent concentration of NaCl. At 0.4 per cent concentration, initial nodulation was not only delayed but the number of nodules and leaves was reduced. These effects were accentuated with an increase in the concentration of the salt and at 0.7 per cent concentration the plants failed to nodulate, indicating that the levels of salinity inhibitory to symbiosis between the legume and the Rhizobium are different from those inhibitory to the growth of individual symbionts.

839.
SUBBA RAO, N.S.; LAKSHMI KUMARI, M.; SINGH, C.S. and BISWAS, A. (1974) Salinity and alkalinity in relation to legume Rhizobium symbiosis. Proc. Indian natn. Sci. Acad. 40B : 544-547.

Effect of chlorides of sodium, potassium and magnesium and carbonate and biocarbonate of sodium, on bacterial (R. meliloti) and host (lucerne) components of a symbiotic system was studied. While R. meliloti could tolerate NaC_l levels upto 3 per cent, seeds of lucerne could not germinate beyond 1.5 per cent. Beyond 0.7 per cent nodulation of lucerne seedlings with R. meliloti was inhibited under aseptic conditions. Almost similar results were obtained with KCl and $MgCl_2$.

The threshold of tolerance for carbonate and biocarbonate was much lower than in the case of chlorides. Nodulation was inhibited at 0.1 to 0.2 per cent levels, though rhizobia could tolerate upto 0.5 per cent and lucerne seedlings upto 0.5-0.75 per cent of carbonate/biocarbonate.

Thus, the degree of salinity/alkalinity conducive for good nodulation were found to be definitely different from the limits of tolerance of Rhizobium and the host to the **respective salts.**

840.
SUBBA RAO, N.S.; MAGU, S.P. and SARMA, K.S.B. (1971) The effect of gum arabic on infection and nodulation of cluster clover by Rhizobium.

Arch. Mikrobiol. 77 : 96-98.

Gum arabic is used to pellet legume seeds with Rhizobium. The gum generally increased the number of infected root hairs of clover seedlings grown on agar, promoted early nodulation and significantly increased the number of leaves during initial nodulation.

841.
SUBHASINI, D. and KAUSHIK, B.D. (1981) Amelioration of sodic soils with green algae. Aust. J. Soil Res., 19 : 361-366.

Algal growth resulted in significant reductions in pH, EC, exchangeable sodium, and improvement in hydraulic conductivity and aggregation status of the soil. There was a significant increase in the total nitrogen content of the soil due to algal growth. Two out of the three inoculated species of algae could establish in the pots along with the indigenous algal flora. Combination of gypsum and algal application were found to have appreciable reclamative properties, and the possibility of using algae as a biological input for the reclamation of sodic soils has been indicated.

842.
SUBRAMANIAN, D. (1967) Biological nitrogen fixation.

J. Indian bot. Soc. 46 : 270-279.

A lecture dealing with the organisms that fix N (Azotobacter, Clostridia, Anabaena, etc.) and with the mechanism of the process.

843.
SUBRAHMANYAN, R. and SAHAY, M.N. (1964) Observations on nitrogen fixation by some blue-green algae and remarks on its potentialities in rice culture. Proc. Indian Acad. Sci. 60B : 145-154.

The net amounts of N fixed after 90-days incubation by algae isolated from rice soil were about 1.61-1.81 g with *Tolypothrix campylonemoides*, 1.26-1.45 g with *Nostoc amplissimum*, 1.2-1.3 g with *N. sphaericum*, and 1.0 - 1.1 g with *Westiella* sp. Potentialities of using blue-green algae as a substitute for N fertilizers for rice under waterlogged conditions are discussed.

844.
SUBRAHMANYAN, R. and SAHAY, M.N. (1965) Observations on nitrogen fixation and organic matter produced by *Anabaena circinales* Rabh. and their significance in rice culture. Proc. Indian Acad. Sci. 61B : 164-169.

In petri-dish experiments, the organic matter produced was 80-92 times the original weight of the inoculated alga, total organic N fixed was 4.7-6.5 mg, and extracellular N was 0.604 mg. In Kjeldahl flasks, the amount of N fixed was 1.2 - 2.5 times the original dry weight of alga inoculated; extracellular N was 0.26 mg.

845.
SUBRAHMANYAN, R.; MANNA, G.B. and PATNAIK, S. (1965) Preliminary observations on the interaction of different rice soil types to inoculation of blue-green algae in relation to rice culture.

Proc. Indian Acad. Sci. 62B : 171-175.

Seven different soils were inoculated with a mixture of the N-fixing species *Nostoc, Anabena* and *Scytonema*, in presence or absence of lime + super + Na_2MoO_4, or treated with $(NH_4)_2SO_4$ at 20-40 kg/ha N. The soils showed differential response to algal inoculation, but all of them were enriched with N and with some growth factor that benefitted the rice crop.

846.
SUBRAHMANYAN, R.; RELWANI, L.L. and MANNA, G.B. (1964) Role of blue-green algae and different methods of partial soil sterilization on rice yield. Proc. Indian Acad. Sci. 60B : 293-297.

Partial sterilization of soil in conjunction with algal inoculation maintained high fertility in plots under rice. Blue-green algae in combination with 500 kg/ha lime, 20 kg P_2O_5 as super and 0.28 kg $NaMoO_4$ were an effective substitute for application of $(NH_4)_2SO_4$ at 20-80 kg N especially in soils of low fertility.

847.
SUBRAMONEY, N. (1950) A new species of Azotobacter isolated from the acid peats (Kari soils) of Travancore-Cochin. Curr. Sci. 19 : 279.

This species, which has been isolated from soils with pH range of 2.5 to 4.5 grows equally well on media supplied with or free from $CaCO_3$.

848.
SUBRAMONEY, N. and ABRAHAM, A. (1963) Incidence of Azotobacter chroococcum in the soils of Laccadive Islands.

Curr. Sci. 32 : 512-513.

A pH range of 6.5 - 7.5 was previously found best for the strain, with pH 5.8 and 9.0 as the limits of tolerance. Soil on the islands contains over 80% $CaCO_3$ and varies in pH from 8 to 9 but A. chroococcum occurred in all soil samples from various islands; this strain when isolated could also grow well in media at pH 4.5.

849.
SUBUDHI, B.P.R. and SINGH, P.K. (1978) Heterocyst spacing in the blue-green alga. Anabaena Azollae.

Curr. Sci. 47 : 510-511.

Heterocyst frequency of Anabaena azollae increased linearly from the apical to the basal 10th leaf in both wild and cultivated Azolla pinnata.

850.
SUD, R.K. and GUPTA, K.G. (1972) On the sensitivity of isolates of Rhizobium spp. and Azotobacter chroococcum to TMTD and its degradation product.

NADDC. Arch. Mikrobiol. 85 : 19-22.

The sensitivity of the bacteria isolated from the rhizospheres of various plants to the seed dressing fungicide tetramethyl thiuram disulphide (TMTD) and to sodium dimethyl dithiocarbamate (NaDDC) was studied at pH values from 5 to 9. A. chroococcum isolates were more sensitive than Rhizobium isolates. Maximum sensitivity occurred at pH 7. Isolates from different plants varied in their sensitivity, and sensitive and resistant isolates differed in their ability to reduce TMTD.

851.
SUD, R.K.; KUMAR, P. and GUPTA, K.G. (1973) On the effect of two herbicides on micro and macro-symbiont.

Zentbl. Bakt. ParasitKde. Abt. II, 128 : 423-425.

The effects of TCA and dalapon on Rhizobium leguminosarum and Pisum sativum were studied in culture and pot experiments. The micro-symbiont was resistant to higher concentrations of these compounds than the macro-symbiont was.

852.
SUDHAKAR BARIK; SIDDARAMAPPA, R.; WAHID, R.A. and SETHUNATHAN, N. (1978) Conversion of p-nitrophenol to 4-nitrocatechol by a Pseudomonas sp.

Antonie Leeuwen. 44 : 171-176

A strain of Pseudomonas sp. ATCC 29354 isolated from parathion-amended flooded soil converted p-nitrophenol to 4-nitrocatechol, which persisted in pure culture. In unsterilized flooded soil, not previously treated with parathion, 4-nitrocatechol was further metabolized by other micro-organisms.

853.
SULOCHANA, C.B. (1952) Soil conditions and root diseases. IV- The effect of microelements on the occurrence of bacteria, actinomycetes and fungi in soil. Proc. Indian Acad. Sci. 36B : 19-33.

Bacterial colonies were considerably inreasing by 50-100 ppm of Mn and B, while they were stimulated by Mo, Li and Zn at lower concentrations. The bacterial population was adversely affected by Ni, Al and Co. Numbers of actinomycetes and fungi were greatly increased by Li and to a lesser degree by Mn, Mo, Co, B and Al, at concentrations of 100-200 ppm. Bacterial numbers increased with increasing time of incubation.

854.
SUNAR, M.S. and CHOHAN, J.S. (1971) Effect of organic and inorganic soil amendments on the rhizosphere mycoflora of groundnut (Arachis hypogaea L.).

Indian J. agric. Sci. 41 : 38-43.

Rhizosphere mycoflora in uninoculated groundnut plants increased with applied gypsum, sulphur and groundnut cake, especially in 60 day old plants. Fewer rhizosphere mycoflora occurred in plants inoculated with Rhizoctonia solani and R. bataticola, Trichoderma viride and T. polysporium occurred in all 60-day old plants and 30 and 45 day old plants treated with gypsum and sulphur but not with groundnut cake. Cephalosporium spp. increased with the age of plants of all treatments.

855.
SUNAR, M.S. and CHOHAN, J.S. (1971) Influence of chemical sprays on the rhizosphere mycoflora of groundnut (Arachis hypogaea L.).

Indian J. agric. Sci. 41 : 29-31.

Sprays of urea, aureofungin and streptocyclin considerably increased the number of rhizosphere mycoflora of groundnut, compared with zinc sulphate sprays. Sprays of aureofungin gave maximum numbers of Aspergillus in 30 day old plants and zinc sulphate the minimum. Colonies of Trichoderma viride and T. polysporium were more apparent after aureofungin and streptocyclin sprays and were negligible after copper-and zinc sulphates in 60 day old plants.

856.
SUNDARA RAO, A.L. (1939) Fixation of nitrogen in soil by the application of pressmud frøm sugar factories. Proc. Soc. Biol. Chem. India 4 : 39-40.

A few results are given showing that application of pressmud to soil brings about fixation of atmospheric N.

857.
SUNDARA RAO, W.V.B. (1964) Soil organic matter and micro-organisms.

Bull. natn. Inst. Sci. India 26 : 221-226.

The possible role of organic matter in supplying energy for the fixation of atmospheric N_2 was investigated. When 5 lb soil mixed with 1 lb fresh cow dung was kept at 13% moisture content, total N rose from 0.13 to 0.142% in 90 days; this increase was equivalent to 160 lb/acre N. The factor limiting N_2 fixation in nature seems to be inadequacy of available energy sources so that bacterial and algal asymbiotic N_2 fixation and possibly photo-chemical fixation may be appreciable in the region of root secretion. In pot tests, inoculation of wheat seed with Azotobacter and cellulose decomposers increased yield and N recovery of the crop.

858.
SUNDARA RAO, W.V.B. (1974) Response of legumes of Rhizobium inoculation. Proc. Indian natn. Sci. Acad. 40B : 604-607.

Recent investigations using acetylene reduction technique by

Hardy *et al.* (1971) showed that nitrogen fixing activity was initiated at 20 to 30 days in soyabeans and continued until senescence at 125 days. About 100 kg N/ha was fixed per season. Indian data showed that the additional amount of nitrogen removed by soyabeans in the inoculated plots over the uninoculated varied from 40 to 120 kg/ha depending upon crop variety, soil and climate. A review of the field experiments conducted in the All Indian Coordinated Pulse Improvement Project showed that location influenced *Rhizobium* strain performance and different varieties of gram and pea responded differently to different strains of *Rhizobium*. The use of peat based cultures of high quality in field experiments gave significant increases in the yields of gram at different centres. However, in some cases neither the cultures nor the nitrogenous fertilisers increased the yield as factors other than nitrogen supply were limiting. In evaluating the efficacy of peat based cultures it is, therefore, necessary to include nitrogenous fertilizers as one of the treatments. Response in the yield of pea to *Rhizobium* inoculation was not adversely affected by the use of pesticides like thiram, captan, malathion and DDT.

859.
SUNDARA RAO, W.V.B. and ISWARAN, V. (1959) Nitrogen fixation by *Azotobacter chroococcum* in culture.

J. Indian Soc. Soil Sci. 7 : 91-95

Azotobacter chroococcum freshly isolated from the soil from Middle Block C_3 fixed 9.3 mgm of N per gm of mannite in nutrient medium containing soil extract. In mixed cultures if repeated transfers were made from one culture solution to another the nitrogen fixing power fell to 4.2. When a loopful of this was transferred from a medium containing soil extract, the nitrogen fixing power increased to 7.8. When a loopful of pure culture from stock culture, maintained by repeated subculturing, was transferred to a liquid medium containing soil extract the amount of nitrogen fixed was 9.5 mgm per gm of mannite, while that transferred to medium containing no soil extract fixed only 3.0 mgms nitrogen showing thereby there was some constituent in soil extract which was necessary to maintain nitrogen fixation at this level.

It was also noticed that the cell-free extract of the organisms did not fix any N which showed that the cellular structure had an important function in the process of N fixation.

The amount of nitrogen fixed from the atmosphere decreased when concentration of NO_3-N was increased. Inhibition of N fixation was almost complete when the concentration of NO_3-N was 1.5 mgm N per 100 cc of the solution.

The amount of N fixed was maximum at the end of 3 weeks. It gradually decreased thereafter. At the end of 6 week period there was practically very little nitrogen in the medium probably due to the autolysis of the organisms and the consequent escape of ammonia formed.

860.
SUNDARA RAO, W.V.B. and SINGH, M.K. (1963) Phosphate dissolving micro-organisms in the soil and rhizosphere. Indian J. agric. Sci. 33 : 272-278.

The P solubilizing powers of bacterial and fungal isolates from alluvial soils, two red-loam acid soils and a hill soil, and from wheat rhizosphere were studied. Acid produced from the carbohydrates influenced P solubility. Inoculation of soil with pure cultures of the organisms increased acid extractable P_2O_5. The organisms identified included Bacillus megatherium, Escherichia freundu, E. intermedia and Bacillus circulans.

861.
SUNDARA RAO, W.V.B. and VENKATARAMAN, K.V. (1963) Studies on the rhizosphere and microorganisms of wheat. Indian J. agric. Sci. 33 : 163-173.

Sampling at tillering and flowering showed significantly higher total bacterial counts in the rhizosphere than in control. Azotobacter was more numerous in the rhizosphere. Application of FYM increased Azotobacter in control and rhizosphere. Symbiotic N-fixing organisms of Rhizobium spp. associated with berseem, soyabean, cowpea and pea could synthesize vitamin B_{12} active substance but little was synthesised by Azotobacter.

862.

SUNDARA RAO, W.V.B., BAJPAI, P.D., SHARMA, J.P. and SUBBIAH, B.V. (1963) Solubilisation of phosphates by phosphorus solubilising organisms using P^{32} as tracer and the influence of seed bacterisation on the uptake by the crop.

J. Indian Sco. Soil Sci. 11 : 209-219

With a view to assess the efficiency of phosphorus solubilising organisms (phosphobacterin-a bacterial fertilizer containing Bacillus megatherium var. phosphaticum obtained from U.S.S.R. and India strain I_1 identified as Bacillus megatherium) in laboratory and pot culture experiments were conducted and phosphate taken up by the wheat crop in a field trial was evaluated.

The extent of soil and fertilizer phosphates solubilised by the introduced organisms was evaluated in laboratory experiments using radiotracer technique also. It was found that in the sterile series both phosphobacterin and Indian strain were equally effective, whereas in the unsterilised series the introduced organism (Russian strain) was more effective than the Indian strain in solubilising phosphate present in the added fertiliser.

Results obtained in laboratory, pot culture, experiments and field trial showed that phosphobacterin was better than Indian strain I_1 specially with respect to bringing into the crop more of fertilizer phosphorus. Phosphate favourably influenced by inoculation with phosphobacterin when farmyard manure was applied as basal dressing.

863.

SUNDARA RAO, W.V.B.; MANN, H.S. and PAUL, N.B. (1963) Bacterial inoculation experiments with special reference to Azotobacter. Indian J. agric. Sci. 33 : 279-290.

In pot experiments, inoculating soil with azotobacter at a level of 10% of soil weight increased grain yield of Eluesine coracana by 57.4% in presence of a basal dressing of berseem hay. Inoculation of wheat seed with azotobacter increased grain yields and N and P uptake.

864.
SUNIL PURI and GROVER, I.S. (1980) Effect of some protein synthesis inhibitors on differentiation of heterocyst in blue-green algae.

Indian J. exp. Biol 18 : 884-886.

The differentiation of heterocysts in Anabaena ARM 314 and Cylindrospermum sp. was suppressed by chloramphenicol, 5-fluorouracil and quinacrine hydrocholoride, inhibitors of protein synthesis. Chloramphenicol was ineffective if added after 6h incubation in basal medium, while 5-fluorouracil and quinacrine were ineffective if added after 2h and 8h of incubation, respectively. Other inhibitors, including cycloheximide, actinomycin-D and nalidixic acid, did not suppress heterocyst differentiation.

865.
SUSHKINA, N.N. (1956) The salt-tolerance of Azotobacter. Mikrobiologia 25 : 35-40 (R).

Azotobacter e.g. A. halophylum often occures in virgin saline soils and in secondarily salinized cultivated soils. The Na, Mg and Ca concentrations limiting N fixation were much higher for A. halophylum than for a species isolated from non-saline podzolized soils. When the two species were cultivated in Ashby's medium and K was added without additional Na, they lost their power to grow. As the ratio of added Na : K was increased from 1 : 3 (weak growth and insignificant fixation) to 3 : 1, growth and fixation increased to an optimum.

866.
SWAMINATHAN, R. and SULLIA, S.B. (1969) Influence of the pesticide malathion on groundnut (Arachis hypogaea L.) microflora.

Curr. Sci. 38 : 282-284.

Spraying with malathion (in dilutions of 1/400 to 1/600) decreased bacterial counts and steadily increased actinomycetes (especially antibacterial species) in the rhizosphere; there was little if any effect on fungal populations.

867.
SWARAJ, K. and GARG, O.P. (1970) The effect of ascorbic acid, when applied to the rooting medium, on nodulation and nitrogen fixation in gram (Cicer arietinum).

Physiologia Pl. 23 : 889-897.

Ascorbic acid applied to plants grown in sand cultures enhanced nodule initiation, increased the number and fresh weight of nodules and delayed their greening. The N content of the plants was 2.5 times that of the control, pod formation started earlier, and pod and seed weights were much higher than in the control. Treatment with 20 mg/l ascorbic acid was more effective than 100 mg/1.

868.
SWARAJ, K. and GARG, O.P. (1978) Inter-relationship between ascrobic acid and structural stability of leghemoglobin in root nodules of vigna sinensis L. and Cicer arietinum L. pp. 164-170. In S.P. Sen, Y.P. Abrol and S.K. Sinha (eds.), Nitrogen Assimilation and Crop Productivity, Associated Publishing Company, New Delhi.

Investigations were carried out on cowpea (Vigna sinensis L.) cv. FS 68 and chickpea (Cicer arietinum L.) cv. C-235 to study the interrelationship between ascrobic acid (AA) and structural stability of leghemoglobin in root nodules. With AA treatments, the leghemoglobin as well as the endogenour AA content of the red and brown nodules were higher than the control. Irrespective of any exogenous treatment. AA content of the red functional nodules was the highest. With the onset of pigment oxidation, the decrease in leghemoglobin content of brown nodules was accompanied by a decline in their AA content which became more precipitous with nodule greening. Although there was a rough parallelism between leghemoglobin relationship was lacking.

869.
SYNGHAL, K.N. (1960) A simpler and more accurate procedure than Iowa method for determining nitrification in soils.

J. Indian Soc. Soil Sci. 8 : 219-222

A procedure for determination of nitrification in soils having the following advantages over the Iowa method has been described:-

a) A large quantity of soil sample is used which is likely to be more representative than the smaller quantity used in Iowa method;

b) the necessity of having suction arrangements is eliminated;

c) the steps of extracting and clarifying the extract are achieved in one operation; and

d) it is simple, rapid and accurate.

870.
TANDON, S.P. and MISHRA, M.M. (1967) Nitrate formation in soil solution by nitrobacter group of bacteria.

Agrochimica 12 : 1-4.

In all the types of soil solution studied (one soil lacking NO_2, and two soils containing NO_2), NO_3 formation by Nitrobacter (isolated from soil by the enrichment and elective culture-method) proceeded rapidly and was sometimes greater than in liquid culture media (sterilized $NaNO_2$ solution, Fred-Davenport medium).

871.
TANDON, S.P. and MISHRA, M.M. (1968) Effect of some organic

acids on nitrification by Nitrobacter agilis.

Zentbl. Bakt. ParasitKde Abt. II, 122 : 401-404.

Maximum nitrite oxidation took place when 50 mg citric acid were present in 50 ml of the liquied culture medium (1.0 g $NaNO_2$ 1.0 g Na_2CO_3 0.5 g k_2SO_4 0.5 g NaCl, 0.3 g $MgSO_4$ traces of $FeSO_4$ 1000 ml distilled water). Fumaric acid had similar effects, but tartaric acid, even in traces, was toxic.

872.
TANDON, S.P. and MISHRA, M.M. (1968) Effect of some polyhydric alcohols on nitrification by Nitrobacter agilis.

Zentbl. Bakt. ParasitKde Abt. II, 122 : 516-519.

Oxidation of NO_2 by N. agilis was enhanced by the presence of 0.4% mannitol (or sorbitol); above this concentration, the alcohols (used here as energy sources) progressively decreased the rate of NO_3 formation to zero at a concentration of about 2%. The longer the bacterium was allowed to remain in contact with mannitol in NO_2-free culture, the smaller its activity became.

873.
TANDON, S.P. and MISHRA, M.M. (1968) Effect of some rare elements on nitrification by nitrate-forming bacteria in soil suspension.

Zentbl. Bakt. ParasitKde Abt. II, 122 : 155-157.

The rate of NO_3 formation by Nitrobacter agilis was accelerated by the presence of $BeSO_4$ $Th(SO_4)_2$ UO_2SO_4 and $ZrCl_4$ in the soil suspension, and was retarded by Tl_2SO_4 and $AuCl_3$.

874.
TANDON, S.P. and MISHRA, M.M. (1968) Effect of some rare elements on nitrification by nitrate-forming bacteria in talc medium.

Agrochimica 12 : 365-368.

Sulphates of indium, thorium, beryllium and uranium and zirconium chlorides enhanced oxidation of nitrate by *Nitrobacter*; thallous sulphate and auric chloride retarded it.

875.
TANDON, S.P. and MISHRA, M.M. (1969) Effect of iron at different pH and temperature on the activity of *Nitrobacter agilis.*

Agrochimica 13 : 323-326.

Fe had no appreciable effect on the rate of nitrite oxidation by N. *agilis* except for a marked delay when the amounts of Fe exceeded 1.0%. At high pH, low temperatures had a favourable effect on nitrification while at low pH the higher temperatures were more favourable. The maximum rate of nitrification occurred at a neutral or slightly alkaline reaction and at temperatures between 20 and 32°C.

876.
TANDON, S.P. and MISHRA, M.M. (1969) Effect of sodium, manganese and zinc on the activity of *Nitrobacter agilis.*

Zentbl. Bakt. ParasitKde Abt. III : 399-402.

Oxidation of NO_2- was promoted by all the elements, provided they were present in the medium in concentrations of (0.05%; higher concentrations than this inhibited NO_2- oxidation, and concentrations of > 1% were toxic.

877.
TARAFDAR, J.C.; ROY, A.B. and MANDAL, A.K. (1981) Enzyme status of some jute-growing soils of West Bengal.

Australian J. Soil Res. 19 : 181-184

The distribution of urease and acid and alkaline phosphatase in some jute-growing soils of West Bengal was investigated.

Urease and alkaline phosphatase activities were significantly and positively correlated with organic carbon, fungal and bacterial populations, whereas acid phosphatase activity was only significantly correlated with organic phosphorus content. None of the soil enzyme activities was correlated significantly with pH, clay content or soil actinomycetes. Urease and alkaline phosphatase were significantly correlated with each other.

878.
TARAR, J.L. and ASHAN JAMAL (1980) Thermoduric properties of soil algae. J. Indian Soc. Soil Sci. 28 : 550-552

Results indicate that blue-green algae generally withstand higher temperatures than the greens, except Chlorococcum humicola. There are indications that resistance to high temperature depends upon some basic features of the organisms and also on the interacting factors like the stability of proteins.

879.
THAKUR, A.K. and NORRIS, R.V. (1928) A biochemical study of some soil fungi with special reference to ammonia production. J. Indian Inst. Sci. 11 : 141-160.

A typical local soil has been examined for its fungus content and some fifty species isolated. The ability of various species to break down cellulose and to liberate ammonia from protein material has been investigated. The mechanism of ammonia production by fungi had been examined and in particular the properties of an amidase isolated from Aspergillus flavus. The reaction is of an autocatalytic nature, the velocity curve first exhibiting a period of lag and then a period of acceleration as ammonia is formed.

880.
THOMAS, P.K.; VENKATARAMAN, C. and VASU, K. (1974) Quality and quantity of peat material reserves in the Nilgiris. Proc. Indian natna. Sci. Acad. 40B : 608-612.

Well decomposed peat material is considered to be rich in plant nutrients, and is a proven carrier for Rhizobium. The

Indian Agricultural Research Institute, New Delhi is using the peat material from the Nilgiris, as a carrier.

The article describes the location, extent and the quality of peat material in the upper plateau of the Nilgiris. Peat material occurs in scattered patches mostly in the concave depressions and invariably associated with small streams. Due to deposition of eroded soil, it is often observed below a mineral soil layer. From the survey conducted, it has been observed that peat material of suitable quality occurs in about 1763 ha which might yield about 3.5 million tonnes of dry peat material.

The data on physical and physico-chemical properties are given and discussed.

881.
TILAK, K.V.B.R. and MURTHY, B.N. (1981) Occurrence of _Azospirillum_ in association with the roots and stems of different cultivars of barley _(Hordeum vulgare)_.

Curr. Sci. 50 : 496-498.

The isolation of _A. brasilense_ from roots and stems, and N fixation in roots isolates and enrichment cultures of stems of 8 barley cv. are described. N fixation by isolates of hulless cv, was markedly higher than by isolates of hulled cv.

882.
TILAK, K.V.B.R. and SAXENA, M.C. (1974) Performance of peat-based single strains of _Rhizobium japonicum_ on soybean _(Glycine max)_. _Proc. Indian natn. Sci. Acad._ 40B : 476-478.

Field experiment was conducted to test the efficiency of various indigenous peat based single strains of _R. japonicum._ It was observed that SB-16 (strain obtained from I.A.R.I. New Delhi) and A-99 (strain obtained from JNKVV, Jabalpur) resulted in higher yields of beans than that under control. Differences between various cultures were, however, not significant. These strains also recorded better nodulation, leghaemoglobin synthesis and bacteroid contents. The viability and respiration rate of rhizobial cells were greater in these strains as compared to that in the others. E-412, UPAU-2 and SB-6 strains resulted in poor yields. The viability and respiration rate of rhizobial cells in these strains were also poor.

883.
TIWARI, D.N. (1978) The heterocysts of the blue-green alga Nostochopsis lobatus : effects of cultural conditions.

New Phytol. 81 : 653-656.

The branched filaments of Nostochopsis lobatus form infrequent intercalary heterocysts in the main trichome while the short lateral branches (1-4 celled) characteristically produce only a terminal heterocysts are not formed but the lateral heterocysts enlarge to assume a balloon-like appearance; they undergc division in situ and liberate 3-5 celled germlings. These grow and produce similar heterocysts and the cycle is repeated until growth ceases. Heterocysts are produced at the terminal ends of the long lateral branches during the later stages of growth in ammonium medium; they are not formed in this position in N-free medium. The possible role of lateral heterocysts in controlling growth and development of long laterals is discussed.

884.
TIWARI, D.N.; MISHRA, A.K. and PANDEY, A.K. (1982) Growth characteristics of the blue-green algae Nostoc linckia and its mutant under reducing conditions.

Comp. Physiol. Ecol. 7 : 199-203.

The nitrogen-fixing alga Nostoc linckia grew better under N_2-free reducing conditions with a slight change in the phycocyanin/Chlaride ratio. The heterocyst formation in the algal filaments under the above conditions was found to alter with respect to the production of paired heterocysts and their frequent germination compared to the normal pattern found in the aerobically grown filaments. Non-heterocysts mutants of the alga neither grew nor produced heterocysts under either conditions and the filaments ultimately converted to unicellular populations compared to the NO_3 grown cultures.

885.
TIWARI, D.P. (1980) A new record of Curvularia gudauskasii Morgan-Jones and Karr from Indian soil. Curr. Sci. 49 : 400-401.

A species of Curvularia was isolated from a soil sample of protected grassland in Jabalpur. Comparative morphology of the present isolate with other species of Curvularia showed that it resembled closely to Curvularia gudauskasii Morgan-Jones and Karr except for much broader, variously shaped conidia. These variations have been treated as an extension of the range of C. gudauskasii and the present isolate is disposed under this species. A review of literature on Indian fungi shows that so far, C. gudauskasii has not been reported from India. The present note, therefore, constitutes the first report. A brief description of the fungus is presented.

886.
TIWARI, D.P., AGARWAL, P.D. and LODH, A.R. (1977) Chaetomium jabalpurense, a new fungus from soil.

Curr. Sci. 46 : 578-579.

A description of this new species isolated from grassland soil is presented.

887.
TIWARI, G.L. (1971) Fixation of elementary nitrogen by Scytonema geitleri Bharadwaja in pure and unialgal cultures.

Curr. Sci. 40 : 330-331.

N fixation by the blue-green alga Scytonema geitleri, isolated from the paddy field soils of Patna, is reported. In pure cultures, the alga fixed 11.25% N (dry weight basis).

888.
TIWARI, V.K. and RAI, B. (1977) Effect of soil burning on

microfungi.

Pl. Soil 47 : 693-697.

Plants residues and dry litter were burnt for 2 hr on a 3 X 3 m plot of alkaline soil at Varanasi, and the mycoflora was monitered during the following 7 months. The number of species and their populations were reduced by burning but returned to normal levels by two months. Species of *Aspergillus* and *Penicillium* colonized the soil first; *Trichoderma viride* appeared after two months and then increased, whereas *Penicillium* decreased with time. *Fusarium oxysporum* appeared after five months.

889.
TREHAN, K. and SINHA, U. (1982) DNA-medicated transformation in *Nostoc muscorum*, a nitrogen fixing cyanobacterium. Australian J. Biol. Sci. 35 : 573-577.

Genetic transformation of an auxotrophic valine requiring marker and a marker with resistance to Pfluorophenylalanine has been demonstrated in *Nostoc muscorum*. Transformation is primarily medated by DNA and is sensitive to ribonuclease and proteniase. The kinetics of the frequency of transformation, which is dependent on the concentration of DNA, suggests a saturation phase. Transformants, though devoid of heterocysts, are able to grow in a medium lacking a combined nitrogen source.

890.
TRIPATHI, S.K. and EDWARD, J.C. (1978) Response of Rhizobium culture inoculation, zinc and molybdenum application on rhizosphere and phyllosphere microbial population of soybean (*Glycine max* Merrill.).

Curr. Sci. 47 : 503-504.

The effects of Zn, Mo and Zn + Mo applied to the soil, and seed inoculation with *Rhizobium*, on the populations of total bacteria, *Azotobacter* and *Rhizobium* in the rhizosphere and phyllosphere of soyabean at the flowering stage were investigated in field plots. Mo and Zn tended to decrease total bacteria in the rhizosphere but increased Azotobacter and Rhizobium; in the phyllosphere, Mo and Zn increased Rhizobium while decreasing other bacteria.

891.
UMA, B.N. and NARASIMHA RAO, P.L. (1959) Actinomycetes. I - Distribution of streptomycetes in Indian soils. Formation of antifungal antibiotics by *Streptomyces champavati* n. sp.

Indian Inst. Sci. Golden Jubilee Res. (1909-1959), pp. 130-141.

Streptomycetes are widely distributed in soils of South India, but are not found in black clayey saline and black cotton soils.

892.
UPADHYAY, R.S. and RAI, B. (1979) Ecological survey of Indian soil fungi with special reference to Aspergilli, Penicillia and *Trihchoderma.*

Rev. Ecol. Biol. Sol. 16 : 39-49.

The distribution of soil fungi in thirty six soil samples collected from different parts of India were studied. More than 70 species representing 31 genera were isolated from different samples during the period September to December 1976. The fungi recorded were mostly deuteromycetous. The most common genera were *Aspergillus, Penicillium, Trichoderma, Spicaria* and *Fusarium. Aspergillus* was the most dominant and most widely distributed soil fungus. *Trichoderma* was recorded very frequently and at high population levels in forest and other soils with high organic matter content and low pH. Aspergilli predominated over Penicillia in cultivated soils, but the reverse was true in soils with high organic matter content. *A. niger, P. chrysogenum* and *T. viride* were the dominant species amongst Aspergilli, Penicillia and Trichoderma respectively.

893.
UPADHYAY, R.S. and RAI, B. (1980) Interrelationship between

Coprinus lagopus and Fusarium udum in soil in relation to competitive saprophytic ability and microbial antagonism. Pl. Soil 56 : 169-173.

Coprinus lagopus was a potent and frequent colonizer of pigeon pea substrates in soil in competition with Fusarium udum. Saprophytic colonization by F. udum of pigeon pea substrate precolonized by C. lagopus, and by C. lagopus on substrate precolonized by F. udum, was inhibited in all inoculum : soil mixtures. Saprophytic colonization of the substrate by F. udum was better at 22°C than at 30°C and that by C. lagopus better at 30°C than at 22°C. Substrate colonization by each fungus was suppressed by the other fungus to an extent which depended upon the temperature. Hyphal parasitism and colony interactions between these fungi and with other saprophytic fungi were investigated. No antagonism was observed between F. udum and C. lagopus. However, these fungi were potent antagonists towards other microfungi. The formation of fruiting bodies by C. lagopus was observed in the laboratory on nutrient media but more frequently on pigeon pea substrates.

894.
UPADHYAYA, S.D.; SIDDIQUI, S.A. and SINGH, V.P. (1981) Seasonal variation in soil respiration of certain tropical grassland communities.

Trop. Ecol. 22 : 157-161.

Four different grassland communities at Ujjain were evaluated for total soil respiration. In the Sehima community respiration rates were higher in the rainy season, a significant reduction occurring in summer. Soil temperature and soil moisture were related linearly to CO_2 evolution.

895.
UPPAL, B.N.; DAJI, J.A. and PATEL, M.K. (1947) Influence of root excretions and germinating seeds on nitrogen fixation by Azotobacter. Proc. Indian Acad. Sci. 25B : 173-177.

An efficient strain of Azotobacter cultured in Ashby's medium fixed 2.10, 0.84, 0.45 and 0.41 mg of N in 20 days in the presence of sterile seedlings of radish, rice, wheat and jowar respectively, compared with 0.48 mg when cultured alone. The addition of 1 gm of sterile soil to the culture raised the amounts of N fixed to 3.17, 1.79, 3.83 and 1.80 mg. None of the seeds tested was capable of fixing elemental N during germination.

896.
VAIDEHI, B.K.; LAKSHMINARASIMHAM, A.V. and RAO, P.R. (1967) Murogenella terrophila from rhizosphere of paddy.

Curr. Sci. 36 : 163.

The isolate differs from the one reported by Goos and Morris (1965) in having wider mycelia, longer conidiophores and conidia. The fungus forms a new record for India.

897.
VARADE, P.A. and PATIL, M.N. (1982) Effect of sodium salts in association with organic manures on nitrification in a vertisol J. Indian Soc. Soil Sci. 30 : 97-98.

Treatment of soil with sodium carbonate or sodium sulphate reduced the nitrification due to increased osmotic tension and toxicity to bacterial activity. Nitrification was stimulated in the presence of FYM and groundnut cake.

898.
VARMA, A.K. and SUBBA RAO, N.S. (1974) influence of foliar sprays of urea on the growth, yield, nodulation and biochemical characteristics of Phaseolus aureus (Baisakhi mung). Proc. Indian natn. Sci. Acad. 40B : 587-598.

Pot trials with Phaseolu's aureus (Baisakhi mung) have shown that foliar feeding of urea was not beneficial to improve the efficiency of legume - Rhizobium symbiosis. Analysis of the rhizosphere microflora of plants sprayed with urea revealed that the number of bacteria and actinomycetes increased with a concomitant reduction in fungal numbers.Likewise the amounts of amino acids in the rhizosphere soils increased due to foliar spray, whereas the amounts of organic acid and sugars decreased. An imbalance in the functioning of root nodules was apparent by the observed reduction in the thio (-SH) and ATP contents of the extracts of nodules from plants sprayed with urea.

899.
VASANTHARAJAN, V.N. (1969) Investigations into the pattern of microbial colonization of early rhizosphere with model systems.

J. Indian Inst. Sci., 51 : 148-154.

Tests with pure cultures of bacteria and fungi (*Pseudomonas, Azotobacter, Arthrobacter, Bacillus, Micrococcus, Aspergillus, Mucor*) and sterilised soil and mulberry seeds showed that soil organis s were able to establish themselves in the rhizosphere better than seed-coat organisms. Bacteria, especially Gram-ve species, were the most efficient colonizers.

900.
VASANTHARAJAN, V.N. and BHAT, J.V. (1968) Interrelations of micro-organisms and mulberry. III. The beneficial influence of certain heterotrophs on the nitrifiers in the rhizosphere.

Pl. Soil 29 : 156-169.

The mulberry rhizosphere stimulated three species of *Pseudomonas* and one each of *Achromobacter* and *Bacillus* capable of degrading methionine and of reversing its inhibitory effect on nitrification in soils. Two of the bacteria were able to form nitrite from methionine. It is suggested that increased nitrifying activity in the mulberry rhizosphere in the presence of methionine results from the activity of these organisms.

901.
VENKATA RAM, C.S. (1960) Foliar application of nutrients and rhizosphere microflora of *Camellia senensis*. Nature. 187 : 621-622.

One year old plants in soil cultures were sprayed with 0.76 m $(NH_4)_2SO_4$ 0.04 m $MgSO_4.7H_2O$, 0.67 m KCl, 1.47 m $NaNO_3$, 0.28 m $Na_2HPO_4.12H_2O$ or 0.83 m urea. All the sprays and $MgSO_4$ and Na_2HPO_4 in particular decreased the number of bacteria in the rhizosphere. The number of fungi was increased by KCl and Na_2HPO_4, decreased by $(NH_4)_2SO_4$ and particularly by $MgSO_4$ and urea, and unaffected by $NaNO_3$. Spraying also affected the genera and species of fungi in the rhizosphere.

902.
VENKATAKRISHNAN, S. and ABROL, I.P. (1981) Amelioration of sodic soil through _Thiobacilli_ inoculation and pyrite application. _J. Indian Soc. Soil Sci._ 20 : 526-529.

Cultures of _Thiobacillus thiooxidans_ and _T. ferrooxidans_ were enriched from an acid mine water. _In vitro_ studies showed that inoculation of pyrite with _T. ferrooxidans_ increased acid soluble iron and sulfate formation several fold. _T. thiooxidans_ enchanced acid and sulphate formation but reduced the soluble iron. Inoculation with both the cultures proved inferior to that with iron oxidisers alone. In column studies with highly sodic soil, surface application of pyrite followed by leaching was more effective in improving the soil than mixing. Inoculation with _T. ferrooxidans_ after surface application of pyrite further enchanced soil improvement while addition of azide inhibited it. The involvement and relevance of bacteria in the oxidation of pyrite and improvement of sodic soils are discussed.

903.
VENKATARAMAN, G.S. (1961) Studies on nitrogen fixation by blue-green algae. II-Nitrogen fixation by _Cylindrospermum sphaerica_ Prasad under various conditions. _Proc. nat. Acad. Sci. India_ 31A : 100-104.

C. sphaerica was isolated from soils. The total amount of N fixed increased during 55 days of study and a considerable amount of extracellular combined N was liberated in the early stages. Growth and N-fixing capacity were poor in darkness but were somewhat improved by glucose. Growth was inhibited at pH values below 6.5 and (with N fixation) was maximal at pH 7.5. In media with relatively high concentrations of added N, the production of extracellular combined N was significantly reduced.

904.
VENKATARAMAN, G.S. (1972) Algal Biofertilizers and Rice Cultivation. Today & Tomorrow's Printers and Publishers, New Delhi, p 1-75.

Historically speaking, most of the earlier interest on the utilization of blue-green algae as a biofertilizer, has been of a very academic nature. Either it has been considered with such a cloistered curiosity as not to permit any extension of its application potential or treated with such an over-enthuastic exaggeration as to grossly oversimplify it.

Based on conservative estimate of the total in situ nitrogen fixation by algae as between 20 to 40 kg per hectare, their importance as a biological nitrogen contributor was considered negligible, though not irrelevant. The whole question of using algae as a biological input, therefore, called for a systematic reexamination in the light of recent agricultural developments. The response of the high yielding rice varieties to algal inoculation, the biological activity of these organisms in presence of high levels of nitrogenous fertilizers and powerful agricultural chemicals and the economics of the preparation and application of the seeding material are some of the aspects that are dealt within this volume.

905.
VENKATARAMAN, G.S. (1978) Blue-green algae as a biological nitrogen input in rice cultivation. pp. 132-141. In S.P. Sen, Y.P. Abrol and S.K. Sinha (eds.), Nitrogen Assimilation and Crop Productivity, Associated Publishing Company, New Delhi.

One of the effective nitrogen fixing biological systems in the rice fields involves certain species of blue-green algae which, expressed on a per hectare basis, contribute about 30-40 kg nitrogen/growth cycle, depending on the ecological conditions. No depression of nitrogenase synthesis occurs with the addition of ammonium-nitrogen below 40 ppm. Since the 40 ppm (120 kg NH_4-N/h) threshold for ammonium-nitrogen is a fairly high one, the risk of repression of nitrogenase could be over looked. Using ^{15}N, evidences have been obtained for a transfer of the fixed nitrogen from these algae to the rice plants. Extensive field trials show that the application of specific strains of these blue-green algae to the rice fields results in a saving of about 30% in the chemical nitrogen fertilizer without impairing the rice yield. Since these algae are living systems, this technology holds enormous potential for the developing nations in the Tropics. Scientists are conducting similar research in Philippines, Egypt, Japan, Morrocco, Senegal. Results generally agree with those from IARI.

906.
VENKATARAMAN, G.S. (1981) Blue-green algae : a possible remedy to nitrogen scarcity.

Curr. Sci. 50 : 253-256.

The technique of growing blue-green algae to supply 20-30 kg/ha of biologically-fixed N for application to rice is described.

907.
VENKATARAMAN, G.S. (1981) Blue-green algae for rice production - a manual for its promotion.

FAO Soils Bull. No. 46, 102 pp.

Discusses the distribution and succession of blue-green algae in rice fields, algal nitrogen fixation, methods for its assay, effect of algal inoculation on soil properties, effect of pesticides on blue-green algae, crop-algal associations in relation to crop growth, establishment of blue-green algae in rice soils, algal application and rice yield, production of algae and its economics.

908.
VENKATARAMAN, G.S. (1982) Nitrogen fixation by blue-green algae and its economic importance.

Trans, 12th int. Congr. Soil Sci. 1 : 69-82.

At a time when prices of fossil fuel-based commercial nitrogen fertilisers are rising sharply, limiting the exploitation of the genetic potential for crop yield, improved use of biological nitrogen fixation could be the key to maximising ecological benefits and minimizing environmental hazards. Blue-green algae, constituting a photo-autotrophic nitrogen fixing system, offer immense scope for introducing a combination of a linear and cyclic fertilisation of the soil with nitrogen. The present review deals with various aspects of blue-green algae : the distribution, ecophysiology, nitrogen fixation, pesticide tolerance and economic importance in crop production systems.

909.
VENKATARAMAN, G.S. and GOYAL, S.K. (1969) Some observations on the effect of nitrogen fixing blue-green algae on certain crop plants.

Mikrobiologia 38 : 709-713.

Applying a mixture of Aulosira fertilissima, Tolypothrix tenuis, Cylindrospermum musciola and Nostoc sp. to the soil at 1 kg/ha increased the grain yields of high-yielding rice varieties by 11.8-27. 5%. In pot experiments, different forms of N fertilizer gave yield responses on algae-treated soil ranging from an augmentation in the case of $(NH_4)_2SO_4$ to a depression with $NH_4H_2PO_4$. The effects of partial sterilization and P fertilizing on algae-treated and untreated soil are compared. P was more effective when applied during the 15 days before transplanting than at the time of transplanting. Dry-matter yields of paprike (Capsicum annuum) and lettuce were increased by treating the soil with Cladothrix anomala.

910.
VENKATARAMANAN, S. and KANNAIYAN, S. (1984) Influence of butachlor on the growth and ammonia assimilating enzymes of Azolla.

Curr. Sci. 53 : 548-549.

Results show that Butachlor not only controls the weeds but also the growth of Azolla. In general the ammonia assimilating enzymes viz GHD, GS and GOGAT have been considerably inhibited at higher concentrations. The inhibition has been found to be significant from 100 to 200 ppm levels and maximum at 200 ppm level.

911.
VENKATARAMANAN, S. and KANNAIYAN, S. (1984) Inhibitory effect of fertilizer nitrogen on the growth of Azolla pinnata. Sci Cult. 50 : 227.

All the levels of fertilizer nitrogen (urea) have recorded significant reduction in the growth of Azolla. Increase in the levels of fertilizer nitrogen has proportionately decreased the growth of Azolla.

912.
VENKATESWARLU, B. and AGGARWAL, R.K. (1980) Influence of different management practicies on microorganisms of a desertic soil. Indian J. Microbiol. 20 : 149-151.

Soil samples (0-23 cm) were collected from under the canopies of established trees (Prosopis cineraria, P. juliflora and Albizzia lebbek), grasslands (Cenchrus ciliaris and Lasiurus sindicus), also from cultivated land and from a bare site kept fallow since 1962. The soil was loamy sand of pH 8.0. Populations of bacteria, actinomycetes and fungi were lower than have been reported for other tropical soils. Among the different managements, grasslands supported the highest, and fallow land the lowest population. Under L. sindicus soil contained higher populations than under C. coliaris. Under P. cineraria populations were higher than under A. lebbek and P. juliflora. Numbers of nitrifying bacteria (Nitrosomonas and Nitrobacter) followed the same pattern as the heterotrophic microflora in all the systems. There was no significant qualitative difference in the composition of the fungal microflora under the different treatments.

913.
VENKATESWARLU, B. and RAO, A.V. (1984) Role of *Azospirillum* in the nitrogen economy of pearl millet under desertic conditions. *Golden Jubilee Seminar on Soil Resources and Productivity Management*, Indian Society of Soil Science, New Delhi 3 : 56.

Azospirillum forms a close association with several varieties of pearl millet grown on the desert soils of the country. Many strains of this bacterium were isolated both from the ecto and endo-rhizosphere of pearl millet roots and identified as *A. brasilense*. Inoculation with this organism was found to enhance the growth and dry matter production in pearl millet in Leonard's jars, pot culture and field experiments. It increased the formation of root hairs and lateral roots in aseptically grown seedlings and resulted in higher root biomass of plants grown in soil. However, there was no increase in nitrogenase activity. The improvement in plant growth observed with different *Azospirillum* strains was not related to their nitrogen fixing capacity but to the phytohormone synthesis.

914.
VERMA, J.S. (1982) Blue-green algae in nitrogen economy of agriculture.

Indian Farmer's Digest 15(2) : 9-14.

Results show that the amount of nitrogen fixation is dependent on the growth of algae and fixation varies from 4 to 50 kg N/ha. The growth of BGA may be encouraged by applying phosphate and molybdate to soils where they are found abundantly and algal inoculation programme should be intensified where their natural occurrence is less.

915.
VERMA, L.N. (1984) Nitrogen fixation, relative growth rate and physiological characteristics of *Azolla Pinnata*. *Golden Jubilee Seminar on Soil Resources and Productivity Management*, Indian Society of Soil Science, New Delhi 3 : 59.

Twelve strains of Azolla studied for nitrogen contribution in six months ranged from 66 to 228 kg/ha under greenhouse condition. Fifty kg P/ha splitted equally in two increased Azolla biomass by 15% and nitrogen fixation by 10% as compared with full single dose at optimum environmental conditions.

Thirty-four strains of Azolla pinnata, a waterfern, associated with nitrogen fixing cyanobacterium Ababaena azokkae showed a good degree of variability in Relative Growth Rate (RGR) among the sources. A negative correlation between rise in temperature and RGR existed from 35 to 45°C, however, four strains exhibited thermoinsensitivity to a moderate degree. Trend of physiological behaviours : mortality, sporulation, pigmentation (brown, green, pink), rhizoidal length and overall vigour were critically characterized and statistical significance worked out. High temperature had a more conspicuous detrimental effect on some strains of Azolla.

916.
VERMA, N.D. (1983) Isolation of anaerobic spore-forming bacteria from soil, mud and water sources of Manipur. Indian J. Anim. Sci. 53 : 87-89.

Four hundred and twenty seven samples of soil, water and mud from the vicinity were collected. On the basis of different biological and biochemical tests the 53 spore-forming anaerobic cultures were grouped as : 10 Clostridium perfringens, 34 C. bifermentans and 9 C. tertium.

917.
VERMA, R.K.; VYAS, S.C.; SHASTRY, P.P. and JAIN, A.C. (1979) Laboratory screening of some herbicides against three soil-borne pathogens. Indian Phytopath. 32 : 394-396.

Out of seventeen herbicides screened at 100 and 200 µg ai/ml for fungitoxic effects by using bioassay technique against Rhizoctonia bataticola, Sclerotium rolfsii and Fusarium oxysporum f. ciceri, Benthiocarb, methabenzthiazuron and linuron reduced the radial growth considerably. In atrazine amended medium S. rolfsii did not form sclerotia.

918.
VIMAL, O.P. and SHUKLA, G.C. (1970) Chemical growth inhibitors liberated during decomposition of submerged weeds and their effect on the growth of rice crop.

Indian J. agric. Sci. 40 : 535-545.

Gas production and composition, Fe transformation and water-soluble organic acid contents of submerged soil were determined for 40 days after the addition of 2% by weight of Cannabis sative, Carthamus oxyacanthus, Croton sparsiflorus, Heliotropium eichwaldi, Lantana camara, Trianthema portulacastrum and Xanthium strumarium. The results are considered in relation to rice growth.

919.
VIRAJ BERI and MEELU, O.P. (1983) Integrated use of inorganic and bio-fertilisers in rice.

Fert. News 28(11) : 33-38.

The results of experiments on the effect of blue-green algae and azolla on rice yield under different agro-climatic conditions in India are presented and discussed. The data revealed that beneficial effect of blue-green algae on rice yield was not obtained in all situations. In some situations blue-green algae was found effective at low level of 25 Kg N/ha whereas in other locations it proved beneficial even at 150 Kg N/ha.

Studies on azolla revealed that it multiplied well under eastern India climatic conditions. In north India high temperature in summer and low temperature in winter limited its multiplication and use in agricultural production. Harnessing atmospheric nitrogen through dhaincha (Sesbania aculeata) green manuring in rice has been found very effective in these areas. Problems and prospectus of biofertilisers in India have also been discussed.

920.
VISALAKSHI, A.; NAIR, M.R.G.K. and ALYER, R.S. (1979) Effect of phorate on mineralisation of nitrogen in a red soil. Pesticides

13(3) : 21-23.

In a pot culture experiment using cowpea as the test crop application of phorate granules to the soil at 1, 2 and 3 kg ai/ha increased the NH_4-N formation by 8.84, 40.25 and 40.87% respectively over control. NO_2-N formation was not significantly affected by the insecticide. NO_3-N formation in the soil was increased by 38.0, 5.25 and 11.58% respectively, the increase under 1 kg ai/ha being the only significant one.

921.
VITTAL, B.R.P. (1976) Studies on litter fungi, I. Mycoflora of *Atlantia* and *Gymnosporia* litter.

Proc. Indian Acad. Sci. B 83 : 133-138.

Only a few fungal species were dominant on the litters of *Atlantia monophylla* and *Gymnosporia emarginata*, each litter type having a characteristic range of dominant species. The occurrence of the less common fungal species also tended to vary between the litter types. Most of the species belonged to the Fungi Imperfecti.

922.
VLASSAK, K. and JAIN, M.K. (1978) Nitrogenase (C_2H_2) activity in the rhizosphere of some inulin-containing plants. *In* Environmental Role of Nitrogen-fixing Blue-green Algae and Asymbiotic Bacteria. *Ecological Bulls. (Stockholm)* 26 : 319-324.

The nitrogenase activity of some inulin-containing plants was studied along with the change in their rhizosphere microbial flora. A relation between the number of N-fixing microorganisms and nitrogenase activity was observed, which also resembled the rate of rhizosphere CO_2 production. The plants showed differences in influencing the nitrogen microflora, although they contained a common polysaccharide substance in their roots. The nitrogenase activity was higher with inulin as an energy source than with glucose. *Dahlia pinnata* and *Taraxacum officinale* were found to posses great potentiality in enriching the soil through asymbiotic N fixers, compared to Cichorium intybus and Helianthus annus.

923.
VYAS, N.D. (1953) Effect of some indigenous phosphates on the fixation of atmospheric nitrogen through pea.

J. Indian Soc. Soil Sci. 1 : 41-46.

A pot experiment was carried out to study the effect of application of indigenous sources of phosphatic viz., superphosphate, ground bonemeal, trichinodule and singhbhum phosphate on fixation of atmospheric nitrogen through pea. The results showed no gain in a soil nitrogen where the crop was not fertilized (control), phosphate fertilization caused little increase in soil nitrogen after the harvest of above ground portion. The amount of additional nitrogen fixed by the phosphate treated pea was 23.6, 29.8, 11.4 and 10.7 mgm respectively with superphosphate, bone-meal, trichinodule and singhbhum phosphate. Calculations based on equivalent P_2O_5 basis showed that superphosphate, bone-meal, trichinodule and singbhum phosphate faciliated fixation of 0.794, 0.845, 0.335 and 0.300 kg N/kg of P_2O_5 (average of two years). The results indicated that superphosphate and bone-meal stand nearly on equal footing while the two rock phosphates were equal in their effect and were inferior to former low sources.

924.
VYAS, N.D. and DESAI, J.R. (1953) Effect of different doses of superphosphate on the fixation of atmospheric nitrogen through pea.

J. Indian Soc. Soil Sci. 1 : 32-40.

Effect of different doses of superphosphate on fixation of atmospheric nitrogen through pea crop was investigated. Phosphate application significantly increase the number of nodules even from 3rd week so also the weight of the nodules and quantity of nitrogen fixed. However, the effect of dose variation was observed from 10th week growth. The dry matter yield also increased, but the effect of dose variation could be observed only at the maturity stage. The increases were 82.2 and 37.5 per cent over control by higher and lower doses of superphosphate (120 and 60 kg/ha) respectively. The nitrogen content of the crop improved due to application of superphosphate was visible from the three week growth stage, but the effect of doses was visible at maturity only. Application of 120 kg P_2O_5/ha provided a nitrogen yield of 1035.7 mgm as against 750.6 mgm when the crop was fertilized with 60 kg P_2O_5/ha and 494.8 mgm with unmanured crop (109.3 and 51.7 per cent increase over control). The soil analysis after the harvest of the areal part showed no improvement in the nitrogen content.

925.
VYAS, S.R. and PRASAD, N. (1956) Production of vitamin B_{12} by rhizobia in Gujarat soil.

J. sci. ind. Res. 15 C : 52.

Rhizobia from bean and guwar produced the most vitamin B_{12} and pea Rhizobia the least.

926.
WALTON, J.H. (1931) Influence of alkaline salts on nitrification in some Indian soils. Indian J. agric. Sci., 1 : 480-494.

Effect of sodium carbonate, sodium sulphate and sodium chloride of varying concentration was studied on the nitrification with ammonium sulphate and cake, on four soils, viz. Pusa, Jorhat, Kalo and Chinsurah. In general, nitrification was depressed and delayed, the extent being variable in different soils. About 0.2 per cent concentration of all the salts tended to depress nitrification in majority of the soils, sodium sulphate being least inhibitive.

927.
YADAV, N.K. and VYAS, S.R. (1971) Note on the response of root-nodule rhizobia to saline, alkaline and acid conditions.

Indian J. agric. Sci. 41 : 1123-1125.

<u>Rhizobium</u> strains from <u>Sesbania cannabina, Crotalaria juncea</u> and <u>Glycine max</u> were tested for tolerance to pH and to various concentrations of NaCl, Na_2SO_4 $NaHCO_3$ $MgCl_2$ KCl and K_2SO_4.

928.
YADAV, N.K. and VYAS, S.R. (1971) Response of root nodule rhizobia to saline, alkaline and acid conditions.

<u>Indian J. agric. Sci.</u> 41 : 875-881.

Four <u>Rhizobium</u> strains specific to lucerne, 3 to <u>Pisum sativum</u> var. <u>arvense</u> and one each for <u>Phaseolus mungo, P. aureus</u> and <u>P. aconitifolius</u> were grown on mannitol-yeastextract broth containing 0.3% salt as Na, K and Mg sulphates and chlorides or as $NaHCO_3$ and on broth adjusted to pH 3-10. All <u>Rhizobium</u> strains were inhibited by 0.4-0.6% $NaHCO_3$. Resistance to NaCl varied from zero in Ps-1 pea strain to complete in M-1 lucerne, <u>Phaseolus aureus</u> and Ps-47 pea strain. Na_2SO_4 had similar but generally less toxic effects than Na_2Cl. Inhibiton by KCl and K_2SO_4 was similar to inhibiton by NaCl and Na_2SO_4 respectively. $MgCl_2$ was generally toxic above 1.0-1.8% but $MgSO_4$ inhibited only the M-2 lucerne strain. The critical pH range was 3.5-5.0 for the lucerne strains and 3.5-4.0 for the other strains. Al the strains survived at pH 10 although growth was reduced in some strains.

929.
YADAV, N.K. and VYAS, S.R. (1973) Salt and pH tolerance of rhizobia.

<u>Folia Microbiol.</u> 18 : 242-247.

Six <u>Rhizobium</u> strains from <u>Glycine max</u>, five from <u>Crotalaria juncea</u> and one from <u>Sesbania cannabina</u> were grown on agar with yeast extract and mannitol containing 0.2-3.0% Na, K and Mg chlorides and sulphates and $NaHCO_3$ and at pH 3-10. Chlorides were more toxic than sulphates. $MgSO_4$ had a stimulatory effect at 1.0%. 0.2-0.8% $NaHCO_3$ was lethal to all <u>Rhizobium</u> strains tested. The <u>S. cannabina</u> strain was the most resistant to salt and alkalinity. pH 10 did not inhibit the growth of any strain but the strains were sensitive to pH 3.5-4.0.

930.
YADAV, N.K.; ABROL, I.P. and VYAS, S.R. (1970) Note on the effect of gum producing bacteria in soil aggregation. Indian J. agric. Sci., 40 : 623-625.

A study was made on the effect of Azotobacter chrococcum, Derxia sp., Rhizobium sp., from groundnut and an unidentified gum-producing soil isolate on the aggregation of normal and sodic soils of Hissar. Rhizobium sp. from groundnut produced 82 per cent aggregation in the sodic soil but only 12 per cent in the normal soil. Other bacteria produced 4 to 20% aggregation.

931.
ZACHARIAH, S.; SANKARAN, K.V. and LEELAVATHY, K.M. (1981) A new species of Septonema from Indian soil.

Mycologia 73 : 208-210.

Septonema verrucosa, a new species of Hyphomycete isolated from soil in Calicut, India is described.

932.
ALAGAWADI, A.R. and RAMACHANDRA REDDY, T.K. (1985). Effect of trifluralin on Rhizobium and its nodulation on groundnut.

Pesticides 19 (9) : 38-39.

Trifluralin stimulated the pure culture growth of A_1 and TAL-309 at 25 ppm but A_1 was found to be inhibited at 50 and 100 ppm and TAL-309 at 100 ppm only. A_4 and RGP strains of Rhizobia were inhibited at all three concentrations suggesting that these strains are highly sensitive to the herbicides.

933.
ALAGAWADI, A.R. and RAMACHANDRA REDDY, T.K. (1986). Effect of trifluralin on Rhizobium and its nodulation on groundnut.

Pesticides 20 (1) : 27-28.

Experiments were conducted to study the effect of herbicide trifluralin on the growth of Rhizobium and its nodulation on groundnut. It was found that trifluralin at 25 ppm and TAL-309 at 100 ppm inhibited all the four strains of Rhizobia. Only strain A_1 and TAL-309 could nodulate at 250 ppm concentration. However, there was no residual effect of trifluralin applied to soil at one and two Kg/ha after an interval of three months.

934.
AMMANI, K.; VENKATESWARLU, K. and RAO, A.S. (1985) Development of vascular-arbuscular mycorrhizal fungi on an upland rice variety. Curr. Sci. 54 : 1120-1122.

The present investigation revealed that mycorrhizal fungi occur very extensively on rice roots under upland conditions. It is therefore worthwhile to consider the impact of these endophytes on phosphorus nutrition and yield potential of rice.

935.
ANUSUYA, D. (1986) Effect of seed treatment with fungicides on nodulation and growth of groundnut (Arachis hypogaea L.) plants.

Pesticides 20 (6) : 21-22.

Seeds treated with different concentrations of fungicides showed variations in nodulation. Generally at higher concentration (1.0%) there was drastic reduction in nodulation. However at lower concentration (0.01%) stimulatory effect was noticed especially with ceresan and agallol treatment.

936.
ASHA JUWARKAR, REWARI, R.B. and JAIN, J.M. (1986) Survival of Rhizobium in stored soils of India. J. Indian Soc. Soil Sci. 34 : 411-413.

It is known that rhizobia of black gram, cowpea and red gram are found almost in all the soil types of India. But none of the rhizobia specific to these three legume could be recovered from 15 soil samples stored for long periods from 1937 to 1979. In general, rhizobia survived for longer time in black and alluvial soils. Out of 16 strains isolated, 11 were confined to alluvial and black soils, which shows the common occurrence of rhizobia of black gram and cowpea. While in the case of red and related soils, the occurrence and survival of rhizobia in general was rather poor.

937.
CHAHAL, R.S.; KHERA, A.P.; SINGH, J.P.; GUPTA, A.P. and SINGH, B.P. (1984) Efficiency of Rhizobium culture in nitrogen management of legume cereal crop rotation. Bull. Indian Soc. Soil Sci. 13 : 376-382.

In field experiments conducted on Typic Camborthid soils for three years Rhizobium inoculation increased the grain yield of green gram, cowpea and black gram and also $KMnO_4$ oxidisable N in the soil. Additional dose of nitrogenous fertilizer further enhanced the yield and soil nitrogen. An economy of 30 kg/ha in fertilizer nitrogen during rabi (winter) season was observed for wheat. Keeping the land fallow during kharif (monsoon season) though improved the soil fertility, it was not economical.

938.
CHAKRABORTY, S.K.; JANA, B.K. and KHATUA, D.C. (1985) Effect of fungicides on Rhizobia, nodulation and seeding pathogen of legumes.

Pesticides 19 (8) : 74-77.

The two protective fungicides namely Emisar-6 and Karathane were inhibitory to all the test cultures of nodule bacteria, whereas all the cultures were insensitive to Allisan, Brassicol, etc.

939.
CHANDEL, A.S.; PANDEY, K.N. and SAXENA, S.C. (1987) Symbiotic nitrogen contribution by nodulated soyabean to non-nodulated

soyabean and ragi. Proc. Symp. Alternate Farming Systems, New Delhi, p. 40.

A field experiment was conducted during kharif 1982 and 1983 to find out the effect of nodulated soyabean on the associated non-nodulated soyabean and ragi. Plant dry matter and grain yield of ragi and non-nodulated soyabean significantly increased by nodulated soyabean. The difference between the N-uptake of nodulated and non-nodulated soyabean was 128.1 kg N ha^{-1} showing that this N was symbiotically fixed. Beneficial effects in terms of mineral N equivalents were 41 and 70 kg N ha^{-1} with one and two rows of nodulated soyabean for non-nodulated one, respectively. Whereas corresponding figures for ragi were 53 and 80 kg N ha^{-1}, respectively.

940.
CHANDRASEKKARAN, P.A. and VENKATARAMAN, G.S. (1985) Regulation of nitrogenase activity in Anabaena variabilis. Proc. Indian natn. Sci. Acad. B51 : 78-84.

Nitrogenase activity in Anabaena variabilis was examined with respect to the effects of exogenous nitrogen, amino acids, antibiotics, inhibitors of protein synthesis, analogues of glutamine and chelating agent EDTA. The effect of NO_3^- and NH_4^+ on performed nitrogenase were quite different in that the inhibitory effect of NO_3-N was much less than that of NH_4-N. The inhibitory effect was reversible, with the removal of NO_3^- and NH_4^+ from the medium.

941.
CHAUDHURI, S. and BHAUMICK, G. (1987) Use of vescicular-arguscular mycorrhiza for agroforestry in acid lateritic soil. Proc. Workshop on Agroforestry for Rural Needs (Abstracts) p. 57-58.

Expansion of agriculture to meet the growing needs for food, fodder and fuel, will involve marginal and submarginal lands which are better suited for agroforestry and silvipastoral uses rather than traditional crop based agriculture. The eroded and degraded acid laterite and lateritic soils of the semi-arid tropics (SAT) are marginal lands, deficient in plant nutrients. Phosphorus in such soils is largely fixed as iron and/or aluminium phosphate and is not available to plants. Improved growth response of plants in presence of vesicular-arbuscular mycorrhiza (VAM) is greatest in soils of low fertility status. Significance of introduction of VAM in the management technology of agroforestry, in such soils, being investigated. Results of an exploratory study with four acid, lateritic soils, from the semi-arid treact of West Bengal, are presented.

942.
DADARWAL, K.R., SINDHU, S.S. and BATRA, R. (1985) Ecology of Hup$^+$ *Rhizobium* strains of cowpea miscellany : native frequency and competence. *Arch. Microbiol.* 141 : 255-259.

Rhizobium strains nodulating the summer legume *Vigna unguiculata*, *V. radiata*, *V. mungo* and *Cyamopsis tetragonoloba* and the winter legume *Cicer arietinum* were surveyed in the northern plains of India and screened for hydrogenase activity to determine distribution of *Hup* (uptake dehydrogenase) in the native ecosystem. It was observed that 56% of the *Rhizobium* strains of summer legumes were *Hup*$^+$ whereas those of *C. arietinum* were all *Hup*$^-$. Acetylene reduction activity *in vitro* was observed in most of the *Hup*$^+$ but not in the *Hup*$^-$ of any of the host species. In the summer legumes mixed inoculation with Hup$^+$ and *Hup*$^-$ strains, under sterilized as well as unsterilized soil conditions, showed that the host species were predominantly nodulated by the *Hup*$^+$ strain.

943.
DEY, B.K. and GHOSH, A. (1986) Effect of molybdenum on nitrogen fixation in Rhizosphere soils of rice. *J. Indian Soc. Soil Sci.* 34 : 493-497.

Foliar spraying of molybdenum (20 and 40 g/ha) stimulated proliferation of *Azotobacter*, *Beijerinckia*, *Nostoc*, *Nostochopsis*, *Aulosira*, *Calothrix* and *Tolypothrix* in the rhizosphere soils of rice (cv. IR-8), while application of nitrogen (60 and 120 kg/ha) promoted the growth of *Azomonas* and *Anabaena* in place of *Aulosira* and *Tolypothrix*. Higher doses of molybdenum and nitrogen together stimulated the nitrogen fixing Bacillus. Nitrogen fixation increased with both the nutrients, with molybdenum intensifying the influence of nitrogen. A high degree of correlation was observed between the amount of nitrogen fixed and the number of nitrogen fixing bacteria and blue-green algae.

944.
DEY, B.K. and TILAK, K.V.B.R. (1984) Biological nitrogen fixation as influenced by soil environment. Bull. Indian Soc. Soil Sci. 13 : 30-50.

The review article stresses that non-symbiotic nitrogen fixation is a major factor in the maintenance of soil nitrogen status. Azotobacter, Beijerinckia, Azospirillum and Clostridium are the most agricultural important soil habitants from view point of non-symbiotic nitrogen fixation. Rhizobium constitute the most important symbiotic association with legumes in fixing nitrogen. The effect of various environmental factors on the activity of these micro-organisms has been discussed. The low efficiency of non-symbiotic nitrogen fixers as compared with the symbiotic ones has been attributed to the largely growth bound nature on nitrogen fixation by the former group and to the fluctuating environmental factors. The symbiotic nitrogen fixers can get rid of the hard microbial competition in soil as soon as they get into the host tissue.

945.
DHAR, B.; RAMA KRISHNA, K.; SINGH, B.D. and SINGH, R.M. (1987) Morphology and biological properties of phage RL 4 of Rhizobium leguminosarum SU 391. Indian J. Exp. Biol. 25 : 38-41.

A virulent phage (RL 4), infective on R. leguminosarum SU 391, has been isolated by enrichment of local soil samples. Phage RL 4 has a polyhedral head (68 nm diam) and a short contractile tail (68 nm length) with a distinct tail plate. The lytic cycle of the phage requires Ca^{2+} ions; Ca^{2+} (1mM) considerably promotes phage adsorption and reduced the latent period, but does not affect the brust size (45 PFU/ml). RL

4 has been found to be more sensitive to heat, UV-light and osmotic shock but more resistant to chelating agent shocks than RL 1.

946.
DIKSHIT, S.N.; KHER, Y.G. and SAXENA, M.C. (1986) Role of Azotobacter in nitrogen fixation in the presence of oil cakes and dicalcium phosphate. Curr. Sci. 55 : 346-347.

It was observed that the Azotobacter count and nitrogen fixation increased in the presence of oil cakes. The increase was enhanced further when DCP and DCP and KCl were also present along with GOC or MOC.

947.
DINESH KUMAR (1987) Mycorrhiza in changing farming systems. Proc. Symp. Alternate Farming Systems, New Delhi p. 12-13.

It has been shown that mycorrhizal fungi are able to improve nutrient absorption significantly without addition of fertilizers when non-mycorrhizal plants require addition of more fertilizers for optimum growth. There is ample evidence that mycorrhizae increase growth and yield of most crop plants by biologically alleviating stress of nutrient, toxic heavy metals, disease, nematodes, salt, draught, adverse soil pH, transplant shock etc. They are especially beneficial in greater absorption of relatively immobile soil elements such as phosphorus, zinc and copper. Mycorrhizal fungi improve revegetation of coal spoils, strip mines, waste areas and other disturbed sites. They have been used with advantage in fumigated or chemically treated areas and green houses where their natural population has been destroyed or damaged. The potential use of these fungi as commercial 'biotic fertilizers' is enormous. Techniques for field scale direct inoculations are however yet to be formulated. Considerably more knowledge must be generated by research on the evaluation and establishment of procedures for the use of mycorrhizae in commercial agriculture.

948.
DORAISWAMY, P. and RANGARAJAN, M. (1985) Tolerance of

rhizobium strains from saline-alkaline soils of Tamil Nadu to sodium chloride and pH. Madras agric. J. 72 : 298-300.

The increase in growth of the majority of rhizobial strains was observed upto a concentration of 0.5% sodium chloride after which there was reduction. The high pH levels had inhibitory effect on the growth of the rhizobial strains of both greengram and blackgram.

949.
GARG, A.P.; GANDOTRA, S.; MUKERJI, K.G. and PUGH, G.J.F. (1985) Ecology of keratinophilic fungi. Proc. Indian Acad. Sci. (Pl. Sci.). 94 : 149-163.

This review evaluates the information available on ecological factors governing the distribution and growth of these fungi in soil.

950.
GARG, F.C., GARG, R.P.; KUREJA, K., SINDHU, S.S. and TAURO, P. (1985) Host-dependent expression of uptake hydrogenase in cowpea rhizobia. J. gen. Microbiol. 131 : 93-96.

Five strains of *Rhizobium* formed effective nodules on *Vigna radiata* cv. K-851, *V. unguiculata* cv. N1, *V. mungo* cv. T-9, *Cyamopsis tetragonoloba* cv. FS-277 and *Cajanus cajan* cv. UPAS-120, as determined by acetylene reduction in excised roots and by plant DW. H uptake was observed only in *V. mungo* nodules; however strain M-11 which did not take up H did not show hydrogenase activity. Acetylene reduction by detached nodules of *V. mungo*, but not those of other hosts increased in the presence of hydrogen. It was concluded that expression of uptake hydrogenase was under the control of the host plant.

951.
GHAI, S.K. and THOMAS, G.V. (1986) And now a new biofertilizer.

Intensive Agric. 24 (6) : 15.

The efficiency of Azospirillum ranges between that of Rhizobium and Azotobacter and is effective mainly on cereal grasses like wheat and millets. It has also been observed that VA-micorrhizal plants perform better under drought conditions. VA-micorrhiza could be used for crops like millets, groundnut, etc. which are usually grown under poor moisture and soil conditions.

952.
GUPTA, R.D. and TRIPATHI, B.R. (1986) Nitrogen fixing capacity of asymbiotic bacteria in some soils of North-West Himalayas. J. Indian Soc. Soil Sci. 34 : 264-270.

In 18 soil profiles of Kangra district (HP), lying in the north-west Himalayas representing various bio-, climo- and litho-sequences; soils of the A horizons, on an average, had about 35, 54 and 53% higher counts of Azotobacter, Beijerinckia and Clostridium, respectively, than those of the B horizons. Grassland soil profiles containing appreciable number of Azotobacter and Beijerinckia occur in the hot and less dry foot-hills, and warm and wet mountainous climosequences, which owed their origin to conglomerate/granitic stone and sandstone lithosequences. Clostridium was almost equal in number in grassland and forest land soils of hot sub-humid low hills siwalik climosequences developing from sandstone lithosequences.

The highest population of Azotobacter and Beijerinckia was in the soils with pH range of 6.1 to 7.0 and Clostridium with 7.1 to 8.7. Organic carbon content of the soils had no effect on Azotobacter and Clostridium while the number of Beijerinckia increased with its content. The nitrogen fixing capacity of the isolates varied widely according to the soils and within the soils according to the bio-, climo- and lithosequences.

953.
GUPTA, R.D.; BHARDWAJ, K.K.R.; MARWAH, B.C. and TRIPATHI, B.R. (1986) Occurrence of phosphate dissolving bacteria in some soils of North-West Himalayas under varying biosequence and climosequence. J. Indian Soc. Soil Sci., 34 : 498-504.

The number of phosphate solubilizing bacteria were determined in 42 soil horizons of 18 profiles at different locations in Kangra district of Himachal Pradesh. Considerable inter-horizon and inter-profile variations in their number were noticed. The population of these organisms was higher in A than in B horizons and in majority of the soils it constituted less than 10% of the total bacteria. Organic carbon content of the soils markedly influenced their population size, while pH in the range of 5.5.-8.0 had no significant role. The bio-and climosequences also showed marked effect on their occurrence and abundance. Grass and forest lands had large population than cultivated soils. The bacterial count was higher in soils under mild and moist climates than in profiles of hot and dry tracts. Four species of P-dissolving bacilli were identified, which showed considerable but variable capacity of phosphate solubilization in three soils treated with rockphosphate and incubated after inoculation with the cultures of these bacteria.

954.
GUPTA, R.P.; KALRA, M.S. and BAJAJ, Y.P.S. (1984) Nitrogen fixation in cell cultures of some legumes and non legumes. Indian J. exp. Biol. 22 : 560-563.

Root derived callus cultures of *Vigna radiata* and *Cicer arietinum* and *Oryza sativa* infected with homologous and heterologous strains of *Rhizobium* established symbiotic relationships with respect to nitrogenase activity and the incorporation of bacteria in the callus mass. Nitrogenase activity, which was higher in the legumes than in rice, was promoted by subculture on a medium containing low levels of nitrogen in the absence of 2,4-D. Plant metabolites (succinic acid and pentoses) also promoted nitrogenase activity. it is concluded that (1) compounds required for nitrogenase activity are present in both legumes and rice and (2) the expression of nitrogenase activity in heterologous systems raises and possibility of incorporating or improving nitrogen fixation activity in other crops.

955.
JAYASANKARI, N. and SHANMUGASUNDARAM, S. (1985) Nitrogen fixation by Nostoc strains. Proc. Indian Acad. Sci. (Pl. Sci.), 94 : 59-64.

Data are presented on the nitrogen fixing potential of seven Nostoc strains in the presence and absence of NO_3^- and NH_4^+ and oxygen.

956.
KAKRALIA, O.P.; YADAV, J.P. and MEHROTRA, R.S. (1985). Effect (in vitro) of some fungicides and an antibiotic on Rhizobium leguminosarum.

Pesticides 19(1) : 52-53.

Eight fungicides viz., Ceresan (dry), Brassicol, Difolatan, Hexathir, hexacap, Benlate, Bavistin, Calexin and an antibiotic Agrimycin, at different concentrations, were tested for their effect on Rhizobium leguminosarum (gram isolate). Brassicol, Benlate, Bavistin and Calixin could not inhibit the bacterial growth. Agrimycin was most inhibitory to growth followed by Ceresan, Difolatan, Hexacap and Hexamir.

957.
KALYAN SINGH, SINGH, K.P., SINGH, T.K. and BOHRA, J.S. (1987) Effect of summer cropping treatment with nitrogen and rhizobium on transplanted kharif rice yield. Proc. Symp. Alternate Farming Systems, New Delhi, p. 5.

Field experiments conducted in 1981 and 1982 at the Research Farm, Institute of Agricultural Science, Banaras Hindu University, Varanasi reveal that cropping treatments showed significant residual effect on succeeding rice crop. Plots where summer mung crop was green manured after two pickings recorded higher values of growth and yield attributes and grain yield of rice as compared to summer mung harvested as grain and straw. Among the seeding time treatments, Rhizobium culture significantly improved the growth and yield attributes and grain yield of succeeding rice crop as compared to control and nitrogen application. Nitrogen application improved the growth and yield attributes and grain yield of rice up to 120 kg N/ha.

958.
KAUSHIK, B.D. (1985) Blue green algae in rice cultivation : An evaluation. Proc. Soil Biol. Symp. Hissar, pp 162-271.

Nitrogen-fixing blue-green algae are abundant in paddy soils. The blue-green algal flora of salt affected soils are little different from the neutral or near neutral soils. Most of the Dominant forms in Indian soils have N_2ase activity. *In situ* algal nitrogen fixation varies from 10-80 kg N/ha/annum. However, application of soil based (10 kg/ha) blue-green algae by broadcasting in the paddy fields after 3-7 days of seedling transplantation provides about 30 kg N/ha/cropping season, in areas where chemical nitrogen fertilizers are not used for various reasons. Algal supplementation helps to reduce the nitrogen dose by one third where the nitrogen fertilizers are used. Even at high levels of chemical nitrogen, algal complementation has beneficial effect on crop yield.

959.
KAUSHIK, B.D. and SUBHASHINI, D. (1985) Amelioration of salt-affected soils with blue-green algae. II - Improvement in soil properties. Proc. Indian natn. Sci. Acad., B 51 : 386-389.

Algalization of paddy crop under sodic soils resulted in a marked reduction in soil pH and exchangeable sodium. An increase in soil aggregation upto 112% and a marked improvement in hydraulic conductivity was observed. The content of carbon and nitrogen, and the availability of phosphorus were also increased.

960.
KAUSHIK, P. (1987) Role of mycorrhizae in agroforestry. Proc. Workshop on Agroforestry for Rural Needs (Abstracts) p. 44.

Mycorrhiza is one of the most important aspect of the microbiota of the rhizosphere. It is believed that under agricultural field conditions crops do not, strictly speaking, have roots they have mycorrhiza. Mycorrhizas have been classified variously on the basis of their morphological characteristics and specific host fungus relationship, as general mycorrhizas (vesiculararbuscular mycorrhizas), ectomycorrhizas (ectotrophic), endomycorrhizas (endotrophic), ect-endotrophic, and host specific (ericoid, orchid and sheathing mycorrhizas). They are caused by the association of non-septate fungi, ascomycetes and basidiomycetes; and have been studied from the roots of the forest trees, of wild herbs and shrubs including the orchids, the highly priced horticultural plants, and from roots of a large number of plants of agricultural crops. In the orchids, very inception of the plant from the minute seed, depends on the entry of fungal partner in case of all terrestrial, lithophytic and epiphytic taxa. Mycorrhizas enhance the uptake of mineral nutrients and facilitate roots in the absorption of water.

961.
KINHAL, G.A. (1985) Use of biofertilizers in bamboo plantations. Indian For. 111 : 502-504.

Application of biofertilizers is both cheaper and better compared to application of inorganic fertilizers. But a balanced combination of Suphala and Azotobacter gives a' significant height growth in case of bamboo (Dendrocalamus strictus).

962.
KISHAN SINGH and MOHAN SINGH (1985) Effect of urea spray with insecticide on phyllospheric microorganisms and nitrogen uptake in sugarcane. J. Indian Soc. Soil Sci. 33 : 925-926.

The results of this investigation has shown that spraying of urea with insecticide greatly reduced the population of phyllospheric microorganisms, thereby resulting in higher utilisation of nitrogen applied through the leaves.

963.
KUMAR, S., SINHA, R.N. and MURTUZA, M. (1986) Effect of fungicides on emergence, nodulation and grain yield of chickpea. Sci. Cult. 52 : 91-92.

Results indicate that some seed dressing fungicides could enhance the gereination, nodulation, dry weight of nodules as well as grain yield.

964.
LAKHANPAL, T.N. and KUMAR, S. (1984) Studies on mycorrhiza and mycorrhizohsphere of Picea smithiana. J. Tree Sci. 3: 5-9.

Two types of naturally occurring mycorrhizae were observed in seedlings of Himalayan spruce: a yellowish white type with a poorly developed harting net, and a brown type. Macroscopic and microscopic characters of the two types are listed in a table. Eleven fungi and two basidiomycetous mycelia were isolated from the mycorrhizosphere of fresh and active mycorrhizal roots.

965.
LAKHANPAL, T.N. and SHARMA, D.D. (1987) Mycorrhizal research on conifers in India - Present status. Proc. Workshop on Agroforestry for Rural Needs (Abstracts) p. 48-49.

The conifers usually form ectomycorrhiza. The ectomycorrhizal research in India is still in its infancy. Research on coniferal mycorrhiza was initiated in India by Bakshi in 1957 in F.R.I. Dehradun. He and his associates studied the morphology and structure of ectomycorrhizae in spruce, silver fir, sal, deodar and pines. They tried to isolate mycorrhiza forming fungi in pure culture and to use the culture as inoculum in treeless regios and in the introduction of exotics. They also studied various management practices in relation to development of mycorrhiza. Lakhanpal and his associates at Shimla, have studied morpho-anatomical characters, nutrient status, mycorrhizosphere, effect of water stress and fumigants and that of artificial inoculation with fungal symbiont on the seedling growth of silver fir, spruce and chilgoza pine. Still lot remains to be done on different aspects of mycorrhizal research on different conifers. Presently, there is a great need to study and characterise the mycorrhiza of various conifers and also to study the mycorrhizal associates of these conifers so as to produce inoculum and define methods and conditions to successfully inoculate nursery seedlings.

966.
LALITA BATRA and RAO, D.L.N. (1985). Effect of Rhizobium

inoculation on nodulation and yield of green gram in an alkali soil. J. Indian Soc. Soil Sci. 33 : 177-178.

The results of this investigation indicate that the Rhizobium legume symbiosis is functional and responses can be obtained with inoculation even in such adverse habitats as alkali soil with pH values ranging upto 9.6.

967.
LALITA BATRA and RAO, D.L.N. (1985) Response of chickpea to inoculants in alkali soil.

Indian J. Agron. 30 : 377-378.

Nodulation status, number of nodules and nodule mass/plant were poor in control and with PAU and JNKVV inoculants indicating the inability of these rhizobia to survive in alkali soils. Nodulation status, nodule number and nodule mass/plant were good with the HAU and IARI inoculants showing the superior ability of these strains to survive under alkali stress.

968.
MAHENDRANATH REDDY, T.; GUPTA, A.K. and VARSHNEY, M.L. (1986) The effect of pesticides on plant growth, nodulation, nitrogen fixation and protein content in fresh peas (Pisum sativum L.).

Pesticides 20 (7) : 36-38.

Rhizobium inoculation in biofertilizer treated plots increased nodulation and nitrogen-fixation. Phorate and BPMC had much depressive action on protein content in fresh peas.

969.
MAJUMDAR, S. and CHANDRA, A.K. (1985) UV-repair and mutagenesis in Azotobacter vinelandii I- Repair of UV-induced damage. Zentb. Mikrobiol. 140 : 247-254.

It is proposed that A. vinelandii OP cells possess at least two dark repair pathways. One of them is active immediately after

UV-irradiation and probably acts before the onset of replication. The other operates after a time lag and therefore possibly after the resumption of replication.

970.

MAJUMDAR, S. and CHANDRA, A.K. (1985) UV-repair and mutagenesis in Azotobacter vinelandii. II - Repair and mutagenesis. Zentb. Microbiol. 140 : 255-258.

Greater numbers of mutants were isolated when UV-irradiated cells of Azotobacter vinelandii OP were treated with caffeine for a limited period of time after UV-irradiation. This was encouraging, considering the difficulty in isolating Azotobacter mutants. Post irradiation treatment with acriflavine, however, yielded a smaller number of mutants.

971.

MANDAL, L.N. and CHHONKAR, P.K. (1984) Soil nitrogen - its chemistry and biology. Bull. Indian Soc. Soil Sci. 13: 1-19.

The article reviews the results of researches carried out so far. Mineralisation of soil organic matter is carried out by diverse groups of microorganisms widely varying in their physiological and nutritional requirements. Unlike mineralisation, nitrification is influenced greatly by environmental factors, the most important among them being adequate oxygen supply. Nitrifiers are very sensitive to soil acidity. Nitrifiers being mesophillic, a temperature of 30°C has been widely accepted as optimum. Organic matter serves as a source of energy for the heterotrophic denitrifiers and therefore, denitrification is stimulated by the addition of organic matter. Soils with reaction in the neutral range have greatest denitrifying potential while a low soil pH is inhibitory. Since denitrification is temperature dependent, denitrification loss of fertilizer N under tropical conditions, where temperature conditions are most optimum, assumes greater significance.

972.

MAURYA, B.R. and SANORIA, C.L. (1986) Beneficial effects of co-inoculating chickpea seed with Rhizobium, Azotobacter and Pseudomonas. Indian J. agric. Sci. 56 : 463-466.

In a two year field study, the grain yield, mass of roots and nodules, gain in soil N and nutrient uptake were influenced more by H 45 than by BG 1 strain of Rhizobium. Co-inoculants had a synergistic effect on rhizobia. Azotobacter B4 was inferior to Azotobacter B5. The best combinations were Rhizobium BG 1 + Azotobacter B5 + Pseudomonas and Rhizobium H45 + Azotobacter B5. An additional dose of 22 kg P/ha significantly increased nodulation, root mass, and soil N at harvest, but not yield.

973.
MISHRA, A.K.; PAMPA ROY, S.S.; ROY MAHAPATRA, S.S. and CHANDRA, D. (1985) Low grade pyrites and their possible benefication by Thiobacillus ferrooxidans. Proc. Indian natn. Sci. Acad., B 50 : 519-524.

An iron-oxidizing chemoautotroph, T. ferrooxidans was found to be quite efficient in oxidation of pyrite from phyriteferous shales. The oxidation was negligible in the absence of essential nutrients. The bacterium in association with the shales showed a great potential to recover copper from a copper ore tailings sample. Five different rock phosphate samples were also made vulnerable to bacterial attack with the help of shales. The leached liquor from the shales was utilized for preparation of chemicals such as ammonium sulfate and iron oxide mass. The latter being usefui as a desulfurizing agent. In addition, other possible exploitations of such unutilized low grade pyrites have been discussed.

974.
MISHRA, B.B., NANDA, D.R. and MISRA, B.N. (1985) Reclamation with blue-green algae : Changes in nucleic acids, protein and nitrogen content of algae exposed to solid waste of a chlor-alkali factory. Environ. Pollut. 37 : 97-104.

Solid waste from a chlor-alkali factory was mixed with garden soil to give mixtures containing from 20% to 60% by weight of soil, inoculated with blue-green algae and incubated for two months. DNA, RNA, protein and nitrogen content of the algal material showed significant decreases with increase in time and in solid waste concentration. An increase in RNA/DNA ratio and a decrease in protein/RNA ratio were also observed. Algae accumulated appreciable amounts of mercury from the waste, showing dependence on both concentration and time. Significant correlations were obtained between mercury uptake and concentrations of the waste and between different combinations of algae and mercury uptake.

975.
MISHRA, D.K.; JHA, M. and KUMAR, H.D. (1985) Involvement of calcium in nitrogen fixation by Nostoc linckia. Curr. Sci. 54 : 43-45.

Two possible explanations for the effect of EGTA on nitrogenase activity were examined. First, EGTA may inhibit Ca^{++} stimulated ATPase activity, inhibiting nitrogenase activity. Secondly, addition of EGTA to cultures of Nostoc linckia might disrupt a Ca^{++} dependent mechanism which normally protects nitrogenase from inhibition by oxygen. Because Ca^{++} in association with superoxide dismutase and catalase is particularly associated with nitrogenase within heterocysts, it prevents toxic effects of superoxide.

976.
MISHRA, M.M. (1985) Solubilization of insoluble inorganic phosphates by soil microorganisms - review. Agric. Rev. 6 : 23-32.

The problem of phosphorus nutrition to plants is complex because only 15-25% of added P is utilized by the plants and remaining gets fixed in soil. There are a variety of micro-organisms in soil which can solubilize native soil phosphorus and insoluble inorganic phosphates. Mechanism of solubilization by different micro-organisms and the crop responses on microbial inoculation have been discussed. The effect of phosphorus enriched compost on different crops and mechanism of solubilization of Mussoorie rock phosphate during composting has also been described.

977.
MOHNAN, R. and SHARMA, V.K. (1986) Photosynthetic pigments of the *Azolla* - *Anabaena* association and its free living symbiont. *Curr. Sci.* 55 : 667-668.

Azolla plants are able to utilize combined nitrogen from the culture medium as well as that which is fixed by the symbiont. The results show stimulatory effect of low levels of combined nitrogen on the *Azolla-Anabaena* pigmentation.

978.
MOHAN SINGH and AGRAWAL, M.P. (1985) Nitrogen transformation during sugarcane root decomposition. *J. Indian Soc. Soil Sci.* 33 : 676-678.

The present study has shown that the microbial population dynamics influenced the periodical changes in soil mineral nitrogen content. The lowest NH^{+}_{4} + NO^{-}_{3} concentration in the soil incubated with roots alone, implicated the need for additional dose of nitrogenous fertilizer to sugarcane ratoons.

979.
MORE, B.B.; SHINDE, N.G. and KONDE, B.K. (1986) *In vitro* studies on the effect of weedicides on growth and nitrogen fixing ability of *Azotobacter* of cotton crop.

Pesticides 20 (11) : 45-46.

In vitro studies on the effect of weedicides viz. TOK-E-25 and Ansar-529 on the growth and nitrogen fixing ability of *Azotobacter* of cotton crop were undertaken. The results show that these weedicides had no adverse effect either on growth or on nitrogen fixing ability of *Azotobacter* if used at the recommended rates.

980.
MOTE, U.N., BAPAT, D.R. and KADAM, J.R. (1986) Effect of *Azotobacter* and carbofuran seed treatment on shootfly incidence and yield of winter sorghum. *Indian J. agric. Sci.* 56 : 457-462.

In a 3-year study, when 'CSH 8 R' 'SPV 86' and 'M 35-1' varieties of winter sorghum (*Sorghum bicolor* (Linn) Moench) were treated with carbofuran and *Azotobacter* alone and in combination, the percentage of dead hearts due to shootfly (*Atherigona soccata* Rondani) got reduced to 13.14-22.11% compared with 27.51% in the control. There was an overall increase of 18.14, 26.85 and 9.74% grain yield and 20.82, 21.33 and 9.15% stover yield in populations treated with carbofuran, carbofuran + *Azotobacter* and *Azotobacter*. For *Azotobacter* it is possible to get 54 times higher monetary return spending only Rs. 10/ha.

981.
MUKHOPADHYAY, A.K., GHOSH, R. and SAHA, D. (1985) Influence of nitrification inhibitor (N-serve) and soil moisture on transformation of mineral nitrogen. *J. Indian Soc. Soil Sci.* 33 : 902-904.

The use of N-serve resulted in an increase in the amount of exchangeable and fixed NH_4^+ in the soil under aerobic incubation. Except under waterlogging, N-serve enhanced the amount of fixed NH_4^+ over no N-serve treatment, presumably due to the inhibition of nitrification that raised the amount of exchangeable NH_4^+ in the soil system, there being a dynamic equilibrium existing between exchangeable and fixed NH_4^+ pools.

982.
MURALIKRISHNA, P.V.G. and VENKATESWARLU, K. (1984) Effect of insecticides on soil algal population. *Bull. Environ. Contamin. Toxicol.* 33 : 241-245.

Parathion, carbaryl and endosulfan, when applied at the recommended levels of field application (5 to 10 ppm), were not harmful to the soil algal population. The application of these pesticides to non-flooded soils at the 5 ppm level slightly enhanced the population of algae while higher concentrations resulted in gradual inhibition, especially under flooded conditions. Parathion caused greater inhibition than the other two. The frequency of the algal forms, however, was not changed due to the application of these chemicals except for certain unicellular forms which were inhibited by the addition of parathion to nonflooded soil.

983.
MUTHUKRISHNAN, P.; MAHENDRAN, S. and KALIAPPA, R. (1985) Studies on the effect of biofertilizers in rice. Madras agric. J. 72. 619-621.

Field experiments conducted at Agriculture College and Research Institute, Madurai during 1980 and 1981 to study the effect of biofertilizers alone and in combination with urea in rice, revealed that application of either blue-green algae or Azolla coupled with 75 kg N/ha gave equal yield as that at 100 kg N/ha, indicating a saving of nitrogen to a tune of 25 kg/ha.

984.
MUTHUVEL, P.; SUBRAMANIAN, V. and SIVASAMY, R. (1985) Effect of organic, inorganic and bio-fertilizer on rainfed redgram. Madras agric. J. 72 : 176-177.

Pooled analysis of two seasons data revealed that benefits realised from Rhizobial seed inoculation is comparable to that of application of either 10 tonnes of FYM or N and P at 20 and 40 kg/ha.

985.
NAGARAJAN, R.; KRISHNAMOORTHY, K.K. and MANICKAM, T.S. (1985) Response of bengal gram to application of copper and phosphorus and Rhizobium inoculation. Madras agric. J. 72 : 361-366.

The results revealed that copper at 5 ppm level recorded a higher grain yield. phosphorus at 40 ppm level was found superior. Rhizobium inoculation resulted in an increase of 50.7 per cent grain yield over no-inoculation, besides increasing shoot weight, root weight, nodule weight and total dry matter production. The results indicated the need of Rhizobium inoculation and application of phosphorus and copper for increased production of bengal gram.

986.
NARAYANAN, R. (1984) Mycorrhizal roles in nutrient cycling and site improvement - a review. In : Proc. IUFRO Symp. on Site and Productivity of Fast Growing Plantations (D.C. Grey, A.P.G. Schonau and C.J. Schutz, eds). South African Forest Research Institute, Pretoria, 2 : 957-968.

This review paper present overall information on the role of mycorrhizae in nutrient cycling.

987.
NARENDRA KUMAR, PAREEK, R.P. and RAMESH CHANDRA (1986) Survival of chickpea rhizobium on seed under different soil moisture levels. J. Indian Soc. Soil Sci. 34 : 196-197.

It could be inferred from the study that irrespective of soil moisture, population of inoculated Rhizobium on seed reduced-stremendously within 48 h of sowing. Subsequently the organism multiplies but population does not attain the original cell count. There is possibility of getting moisture stress tolerant strains like G-4 which could comparatively do well under low soil moisture conditions.

988.
PAHWA, M.R. (1986) Response of forage cowpea to phosphate and rhizobium inoculation. Agric. Sci. Digest 6 : 96-97.

Investigation in field was carried out to study the response of cowpea to different levels of phosphate (0, 10, 20, 30, 40, and 60 kg P_2O_5/ha) and inoculation with Rhizobium. Inoculation alone increased both green as well as dry matter yields of cowpea. the improvement was further increased when inoculation was done in the presence of 30 kg P_2O_5/ha.

989.
PALANIAPPAN, S. and BALASUBRAMANIAN, A. (1986) Effect of two pesticides on rhizosphere microflora of cowpea.

Pesticides 20 (1) :19-20.

Application of corbofuran at 2 ppm increased the population of bacteria, fungi and actinomycetes in the rhizosphere, but had a suppressing effect at 5 and 10 ppm. In fluchloralin also 2 ppm application caused stimulation of bacterial and fungal population while actinomycetes remained unaltered. However, the 5 and 10 ppm levels of fluchloralin proved to be deleterious to rhizosphere microflora.

990.
PANDEY, A.K. (1985) Effects of propanil on growth and cell constituents of Nostoc calcicola. pp 157-162. In : Pesticide Biochemistry and Physiology. Kutir Degree College, Kutir-Chakkey (India).

In studies with the cyanobacterium N. calcicola, propanil inhibited heterocyst differentiation and N-fixation, together with a decrease in total protein, nucleic acids and carbohydrate content and a loss of photosynthetic pigments. Chlorophyll - a synthesis was reduced.

991.
PANDHER, M.S.; GUPTA, R.P. and BHUPINDER KAUR (1986) Lentil plenty with Rhizobium.

Intensive Agric. 24 (1) : 21.

Studies conducted on the use of Rhizobium in the lentil showed increased yields (29.8.70.0%). Only the rhizobium specified for lentil should be used for lentil. Rhizobium inoculation should not be mixed with chemical fertilizers.

992.
PAREEK, R.P.; RAMESH CHANDRA and ZAIDI, S.F.A. (1985) Effect of streptomycin on the growth of Rhizobium. Curr. Sci. 54 : 1138-1140.

Data showed that all the mutant types grew slowly as compared to their parent types. The difference in the growth rates of the wild types and the mutants of chickpea Rhizobium in the initial stage of exponential phase was higher. in view of these results there appears a serious danger in the use of antibiotic resistant mutants in ecological studies because such slow growing mutants may be poor competitors for nodule forming sites on roots as well as for the other growth limiting factors.

993.
PATEL, G.S.; LODHA, P.C. and PATEL, I.S. (1985) Compatibility of seed dressing fungicides with Rhizobium sp. and their effect on nodulation, nodule weight and yield of greengram (Vigna radiata Wilc.) in field condition.

Pesticides 19 (8) : 40-41.

The seed dressing with brassical, captan and thiram enhanced nodulation as compared to Rhizobium treated one. The seed treated with captan and dithane M-45 recorded maximum wet and dry weight of nodules as compared to control.

994.
PATEL, H.R.; THAKAR, N.A. and Patel, C.C. (1985) Azolla and mustard cake against nematicides for root-knot management in okra. Madras agric. J. 72 : 593-594.

Results show that prior application of nematicides by 20 days of seeding has no effect on control of root-knot nematodes, while, organic matter like dry azolla and mustard cake have very good responses in root-knot nematode control and in improving plant growth of okra which is a highly susceptible crop to these nematodes.

995.
PATRA, D.D., SACHDEV, M.S. and SUBBIAH, B.V. (1986) Nitrogen uptake and efficiency in wheat-gram and maize-cowpea intercropping system. Fert. News 31(10) : 21-27.

Three greenhouse experiments and one field experiment were conducted taking wheat (Triticum aestivum L.) - gram (Cicer arietinum L.) mixed cropping during Rabi 1980-81, and maize (Zea mays L.) - cowpea (Vigna unguiculata L.) mixed cropping during summer as well as Kharif 1981. The mixed cropping had distinct beneficial effect on per cent N content, dry matter yield and total N uptake of the mixed cereals, whereas in case of mixed legumes although the dry matter yield was not affected, the per cent N content and the total N uptake were decreased. Inoculation of cowpea with Rhizobium resulted in significantly higher dry matter yield and total N uptake of the associated cereal as well as cowpea both under mixed and pure cropping.

996.
PATRA, D.D. and SUBBIAH, B.V. (1987). Nitrogen fixation and transfer of fixed nitrogen to cereal in a maize-cowpea mixed cropping system. Curr. Sci. 56 : 23-25.

If the role of seed N is considered to be insignificant, the accurate estimate of the N fixed by legume and that transferred to the associated cereal would be the apparant estimate less the contribution from the unknown sources, which is a very small quantity. At maturity cowpea contributed about 27% of the total N uptake by maize. In green house conditions cowpea contributed about 32% of the N uptake of maize.

997.
PRASAD, B., PRASAD, R. and PRASAD, J. (1986) Evaluation of nitrification retardation property of non-edible oils and their influence on yield and N uptake by wheat in calcareous soil. J. Indian Soc. Soil Sci. 34 : 281-285.

In a field experiment, the relative efficiency of nitrification inhibition by different non-edible oils namely, neem (Azadirachta indica L.), karanj (Pongamia glabra Vent.), mahua (Bassia latifolia L.), undi (Ricinus communis L.) and ratanjyoti (a non-edible oil plant growing in Maharashtra) and its effect on yield of and nitrogen uptake by wheat was studied. All non-edible oils retarded the conversion of NH_4^+. The accumulation of NH_4^+ increased with time, which resulted in the highest accumulation of NH_4^+ and the lowest formation of nitrate at maturity stage. The highest nitrification inhibition (73.9%) was observed with application of undi oil followed by ratanjyoti oil (67.3%) and karanj oil (64.2%) at tillering stage. These treatments increased yield of and nitrogen uptake by wheat. The highest yield of and nitrogen uptake by wheat was, however, obtained with three-split application of 100 kg N ha^{-1} as urea.

998.
PRASUNAMMA, K. (1986) Intrinsic antibiotic resistance patterns of some rhizobial strains isolated from tree legumes. Curr. Sci. 55 : 671.

MIC values and the patterns of interinsic antibiotic resistance were studies for all the rhizobial strains. Associations between resistance pattern were random within the strains of the same host and among different hosts. Each strain was diverse, showing different patterns of resistance to different antibiotics.

999.
PURUSHOTHAMAN, D.; CHANDRAMOHAN, S. and OBLISAMI, G. (1986) Nitrogenase depressed mutants in Azospirillum. Curr. Sci. 55. 1192-1194.

All the mutant clones selected by the dye reduction method exhibited nitrogenase activity in the presence of ammonium chloride and potassium nitrate which was much lower than it was under elemental nitrogen. The stability of the mutants was checked for fiften generations and the mutants were quite stable.

1000.
RAM. G., CHANDRAKAR, B.S. and KATRE, R.K. (1985) Influence of Azotobacterization in presence of fertilizer nitrogen on the yield of wheat. J. Indian Soc. Soil Sci. 33 : 424-426.

The results obtained leads to the conclusion that azotobacterization can improve the yield of wheat especially in presence of fertilizer nitrogen.

1001.
RAM. G.; CHANDRAKAR, B.S.; SINGH, K. and BEOHAR, B.B. (1986) Double your benefit from blue-green algae.

Intensive Agric. 24 (3) : 7-8.

The use of blue-green algae at 10 kg per hectare without supplementing with fertilizer, can increase the rice yield to the tune of 19% over the control. Further, the residual impact of blue-green algae on gram yields is still appreciably higher.

1002.
RAM, H.; SINGH, B.N. and PRASAD, J. (1985) Seasonal variations in enzymatic activity in cultivated, pond and virgin soil. J. Indian Soc. Soil Sci. 33 : 805-810.

Seasonal variations in enzymatic activities in two surface layers (0-7.5 cm and 7.5-15 cm) were studied in a cultivated, pond and virgin soils. Invertase activity was highest in cultivated soils than in pond and virgin ones. Soil layer at 0-7.5 cm exhibited greater activity than the other one and rainy season was more favourable than summer and winter seasons. Amylase activity was greater in virgin and pond soils and a decrease in it was observed in second layer (7.5-15 cm). Phosphatase activity was maximum in virgin soil followed by pond, upland and transitional cultivated soils. This activity was significantly higher in 7.5-15 cm layer in rainy season.

1003.
RAMASWAMY, P.P.; OBLISAMI, G. and CHELLAMUTHU, S. (1985) Influence of nitrogen, phosphorus and inoculation on nodulation and yield in blackgram (Vigna mungo L.). Madras agric. J. 72 : 658-660.

Grain yield was significantly increased due to rhizobial inoculation.

1004.
ROY MAHAPATRA, S.S., MISHRA, A.K., CHANDRA, D., PANDALAI, H.S., GHOSH, P.K. and BANERJEE, P.K. (1985) Oxidation of pyrite from pyritiferous shales using Thiobacillus thiooxidans. Indian J. Exp. Biol. 23 : 42-47.

An iron oxidising chemoautotroph, T. ferrooxidans was utilized to investigate pyrite oxidation from pyritiferous shales. Pyritiferous shales were found to harbour iron oxidising bacteria which were effective in pyrite oxidation. The oxidation rate depended on the initial acidity and pulp density of the system. The lag observed during pyrite oxidation by the cells grown on soluble ferrous iron could be eliminated by two stage adaptation on pyritiferous shales. Complete bacterial pyrite oxidation could be achieved by pretreating the shales with 2N H_2SO_4. During leaching of low grade pyrite the inhibition or restriction of bacterial activity with respect to toxicity of metal ions or autotoxic materials was also investigated.

1005.
SAHU, G.C. and BEHERA, B. (1984) Contribution of pulse crops to the nitrogen need of succeeding rice crop. Bull. Indian Soc. Soil Sci. 13 : 383-386.

Three pulses *viz*, greengram, blackgram and horsegram were grown during *rabi* (winter) and a rice crop was taken in the same plots in the following *kharif* (monsoon) season. The yield of rice was significantly higher over control (no legume), the highest increase being recorded where blackgram was grown. Assuming that the increased yield was mainly due to pulse cultivation, the nitrogen fixation was calculated to be of the order of 21, 26 and 16 kg/ha due to greengram, blackgram and horsegram, respectively. The plots under pulses recorded higher quantities of total nitrogen. Harvesting pulse crop by cutting rather than by uprooting was observed to benefit the succeeding rice crop to a greater extent.

1006.
SAHUL HAMEED, M.; UDAYASURIYAN, V.; ANTHONI RAJ, S. and OBLISAMI, G. (1985) Comparative efficiency of peat, lignite and soil farmyard manure mixture as carrier for rhizobium. Madras agric. J. 72 : 178-180.

Results show that *Rhizobia* survived longer in carrier containing glutamic acid. Soil-farmyard manure (1:1) mixture enriched with 100 ppm glutamic acid might serve as a carrier equivalent to peat where peat deposits are lacking and at the same time involves lesser cost in inoculant production.

1007.
SARKAR, A.K.; MALLIK, M.K. and PRASAD, S.S. (1985) Studies on the combined use of bio-and chemical fertilizers on yield and nitrogen uptake in a rice-wheat sequence. J. Indian Soc. Soil Sci. 33 : 187-189.

Results show that algalization could take care of the initial nitrogen requirement of rice crop and may be used in place of any basal application of fertilizer nitrogen.

1008.
SAVALGI, V.P. (1986) Azolla for rice.

Intensive Agric. 24 (4) : 14.

Azolla can be used either as green manure or as a dual crop with rice. After its application to the soil in flooded condition, the fern decomposes quickly thus releasing about 80 per cent of its nitrogen to the growing crop of rice in three to four weeks. Azolla attains full growth after 20 to 40 days depending on the quantity of inoculum.

1009.

SHAH, J.P. and JOSHI, H.U. (1986) Biofertilisers for improving production of cereal crops. Fert. News 30(10) : 75-77.

Biofertilisers are cheaper and renewable and contribute to the development of strategies which do not lead to exponential rise in consumption of non-renewable forms of energy. Considering the importance of biofertilisers, GSFC has taken lead in establishing research, development and production complex to update the technologies and smoothen the bottlenecks so as to make it successful and adoptable by the farmers. Priority has been given to develop efficient strains of cereal crops. Work has been taken up for legumes and oilseeds.

1010.
SHARMA, C.P., GUPTA, B.R. and BAJPAI, P.D. (1986) Residual effect of leguminous crops on some chemical and microbiological

properties of soil. J. Indian Soc. Soil Sci. 34 : 206-208.

Results show that the soils under all the legumes harboured higher microbiological load except *Azotobacter* as compared to those under the cereals or fallow. The higher population of bacteria, actinomycetes and fungi under legumes might be due to the preferential stimulation exerted by the excretory products of their roots. the reduced population of *Azotobacter* in legume cropped soil was probably due to higher availability of nitrogen.

1011.
SHARMA, M.L. and VIJAY SINGH (1986) Relative efficiency of fertilizer nitrogen and bio-fertilizers in rice under irrigated conditions. Indian J. Agron. 31 : 69-74.

Application of 60 kg N/ha through chemical fertilizer with blue-green algae inoculation was better than 60 kg N/ha - without the blue green algae inoculation, both in terms of total production and net profit. The highest content of total N in plants at the panicle initiation stage and crude protein content in paddy grain were recorded with application of 60 kg N/ha with blue green algae inoculation.

1012.
SINDHU, S.S. and DADARWAL, K.R. (1985) Protoplast formation and regeneration in *Rhizobium* and *Azospirillum*. Curr. Sci. 54 : 344-346.

The sucrose-cation-lysozyme (A) and tris-sucrose-EDTA-lysozyme (B) methods both resulted in protoplast formation in three *Rhizobium* strains and in *Azospirillum brasilense*.

1013.
SINGH, A.L. and SINGH, P.K. (1986) Effect of carbofuran application on the growth and nitrogen fixation of *Azolla* during its cultivation with rice. Curr. Sci. 55 : 1102-1103.

Results show that mixing carbofuran with *Azolla* inoculum enhanced the growth and N_2-fixation of *Azolla* and also benefitted the rice crop.

1014.
SINGH, D.P. and SINGH, P.K. (1986) Growth and nitrogenase activity in Azolla species. Indian J. Plant Physiol. 29 : 99-103.

Growth and nitrogenase activity were determined in thirteen collections of Azolla; eleven from *A. pinnata* and one each from *A. mexicana* and *A. filiculoides* under identical conditions in mineral medium. Many isolates of *A. pinnata*, and *A. filiculodes* grew better than *A. mexicana* and among different collections of *A. pinnata*, isolates of Thailand, Bangladesh and Vietnam produced more fresh and dry matter while the minimum growth was recorded in Nepal isolate. Though *A. mexicana* showed poor growth, however, nitrogenase activity was maximum in this isolate whereas, *A. filiculoides* shcwed a low level of nitrogenase activity. Among *A. pinnata* isolates, Bangladesh isolate exhibited highest nitrogenase activity and the lowest was recorded in Indonesia isolate.

1015.
SINGH, R.S. and GUPTA, N. (1985) Effect of soil type, phosphorus application and inoculation on yield and phosphorus uptake in bengal gram. J. Indian Soc. Soil Sci. 33 : 679-682.

Results show that inoculation of seeds of bengal gram with efficient *Rhizobium* strain could increase the yield substantially in different soil types. Bengal gram may perform better in light textured soils than in heavier ones, especially with phosphatic fertilizers and inoculation.

1016.
SOMANI, L.L. (1984) Use of low grade pyrites as an amendment for alkali soils and to improve soil fertility - A review. Fert. News 29 (7) : 13-27.

The efficiency of pyrites in reclaiming alkali soils and as a source of fertilizer S depends upon its oxidation by *Thiobacillus*. The optimum temperature is 25-40°C, moisture content of 60% of water holding capacity and pH below 5.0 for the activity of *Thiobacilli*. It is always advisable to inoculate the soil with a mixed culture of *Thiobacilli* in order to ensure fast oxidation of pyrite.

1017.
SOMANI, L.L. (1986) Use of pyrites as a soil amendment. *Sulphur in Agriculture* 10 : 16-20.

The article stresses that for all practical purposes, oxidation of pyrites in soil is largely a biological process. The oxidation is governed by the substrate the pyrites, the microorganisms and the environment. The pyrite oxidising *Thiobacilli* have shown remarkable tolerance to high concentrations of such heavy metals. *Thiobacilli* could work under conditions of 4°--50°C temperature, 40-60 per cent moisture and pH range of 2.0 to 9.6. *Thiobacillus movellus* is capable of tolerating very high concentration of Na in its environment. The role of *Thiobacilli* is in accelerating the otherwise very slow oxidation of pyrites.

1018.
SRIVASTAVA, G.P. and VERMA, U.K. (1986) Response of pigeonpea to phosphorus and *Rhizobium* inoculation. *Indian J. Agron.* 31 : 131-134.

An experiment was conducted at the Ranchi agricultural college farm, Kanke during *Kharif* 1979-80 and 1980-81. The results revealed that phosphorus up to 60 kg P_2O_5/ha markedly increased the grain yield of pigeonpea. *Rhizobium* inoculation alone or in combination with nitrogen significantly increased the grain yield.

1019.
SUBBA RAO, N.S. (1983) Nitrogen-fixing bacteria associated with plantation and orchard plants. *Can. J. Microbiol.* 29 : 863-866.

Azotobacter and *Beijerinckia* were isolated from rhizosphere soil sampled from 21 plantation and orchard plant species in southern India. *Azospirillum* was isolated from washed and surface sterilized roots of 14 of these plant species: eight isolates were A. brasilense and six were A. lipoferum. Micro-kjeldahl analysis and acetylene reduction procedures demonstrated that all the bacterial isolates fixed nitrogen in pure culture on N-free media. Motile bacteria capable of reducing triphenyltetrazolium chloride were more abundant in xylem cells than in cortical cells of roots.

1020.
SUBBA RAO, N.S. (1984) Interaction of nitrogen fixing micro-organisms with other soil microorganisms, pp 37-64. In : Current Developments in Biological Nitrogen Fixation (N.S. Subba Rao, ed.) Edward Arnold Publisher Ltd., London.

Some ecological aspects of microbial interactions in soil with reference to nitrogen-fixing bacteria are reviewed and the importance of these studies in inoculant preparation and application is indicated.

1021.
TAURO, P. (1984) Role of micro-organisms in the systems approach to plant nutrients. In : FAI seminar 1983 on systems Approach to Fertiliser Industry. FAI, New Delhi 2 : 1-8.

Rhizobial inoculation under Indian condition is discussed.

1022.
THAKUR, R.C. and NEGI, S.C. (1985) Effect of fertilizers and rhizobium inoculation in blackgram.
Indian J. Agron. 30 : 501-504.

Field experiments were conducted to evaluate responses of blackgram (Vigna mungo (Linn.) Hepper) to nitrogen, phosphorus and rhizobium inoculation. Results show that Rhizobium inoculation of seed increased the grain and straw yields.

1023.
TILAK, K.V.B.R.; SCHNEINDER, K. and SCHLEGEL, H.G. (1984) Autotrophic growth of strains of Rhizobium and properties of isolated hydrogenase. Curr. Microb. 10 : 49-52.

Four strains of Rhizobium were demonstrated to grow litho-autotrophically with molecular hydrogen as a sole electron donor and with ammonium or with N_2 as nitrogen source. All of them showed ribulose-1, 5-bisphosphate carboxylase activity and hydrogenase (H_2-uptake) activity with methylene blue and oxygen as electron acceptors.

1024.
TIWARI, G.K. and STIVASTVA, J.S.L. (1986) Response of paddy crop to algalization. Agric. Sci. Digest 6 : 96-100.

A field trial was conducted to assess the effect of blue green algae biofertilizer, a source of nitrogen, on the yield of paddy crop. It was observed that yield in presence of 30 kg N/ha and algae was comparable to 90 kg N/ha alone. Similarly, algae in conjunction with 60 kg N/ha was found to be comparable to 120 kg N/ha. hence, accordingly poor farmers can be benefitted who cannot afford to purchase and apply higher dose of nitrogen as a chemical fertilizer.

1025.
UDAYASURIYAN, V.; PRABAKARAN, J. and OBLISAMI, G. (1985) A simple technique for Rhizobium plant infection test. Curr. Sci. 54 : 515-516.

A simple technique has been described. The advantage of lower cost and labour observed in plastic pouches are retained in this technique besides an added advantage of protection from frequent contamination due to its compactness.

1026.
UDAYASURIYAN, V.; SAHUL HAMEED, M.; ANTHONI RAJ, S. and OBLISAMI, G. (1985) Symbiotic promiscuity of cowpea Rhizobium with soyabean. Madras agric. J. 72 : 230-231.

The dry matter production of soyabean was significantly increased due to inoculation with all the thirteen rhizobial strains of cowpea, even with isolates that showed no nodulation in plant infection test. The nitrogen fixing potential and plant growth augmentation effect of certain cowpea group Rhizobium strain might be greater in soyabean than soyabean specific strain under local soil conditions.

1027.
VAISHAMPAYAN, A. (1983) Vanadium as a trace element in the blue-green alga. New Phytol. 95 : 55-60.

Molybdenum, applied to laboratory media in the form of sodium molybdate at a concentration of 0.0177 ppm, was required for nitrogenase and nitrate reductase activities of these enzymes, and the growth of Mo-containing, cultures were significantly enhanced by the addition of 0.0125 ppm of vanadium, applied in the form of sodium vanadate. At concentrations above 0.0125 ppm vanadate addition inhibited nitrogenase and nitrate reductase activities and ultimately led to a retardation of growth on N_2 and NO_3^-. Molybdenum and/or vanadate had no effect on the growth of N. muscorum with NO_2^- as the nitrogen source.

1028.
VAISHAMPAYAN, A. (1984) Biological effects of a herbicide on a nitrogen fixing cyanobacterium (blue-green alga) : an attempt for introducing herbicide-resistance. New Phytol. 96 : 7-11.

The inhibitory effects of monuron on growth and heterocyst formation in Nostoc muscorum were glucose-reversible in both N-free and nitrate-containing media. A spontaneous mutant of N. muscorum, resistant to 0.36 mM monuron as applied to rice field in North India, was isolated and found to have no requirement for an organic C source for its normal physiology and was capable of resisting 0.44 mM monuron. Monuron was considered to be unable to enter the mutant line.

1029.
VAISHAMPAYAN, A. (1984) Copper-iron interactions in the N_2-fixing cyanobacterium Nostoc muscorum. J. Pl. Nutr. 7 : 567-573.

The N_2 fixing cyanobacterium Nostoc muscorum was found

to be sensitive to $CuSO_4$ and $CuCl_2$. Copper salt at lower concentrations interfered with the nitrogenase activity and growth of the organism in minimal medium, but its inhibitory effects were reversed on exogenous Fe supplementation in the form of $FeSO_4$ or $FeCl_3$. The effects of Cu may be due to Cu-induced Fe-deficiency in N. muscorum resulting from the inactivation of the Fe-Mo cofactor responsible for cyanobacterial N_2 fixation and growth. Higher concentrations of Cu have been found toxic and lytic to the organism, thus leaving no scope of protection by exogenous Fe. Copper had no pronounced effect on heterocyst forming capacity or on NH_4^+ assimilation in the cyanobacterium.

1030.
VAISHAMPAYAN, A. and HEMANTARANJAN, A. (1984) Physiological significance of vanadium uptake during N_2 and NO_3^- metabolism in various strains of a N_2-fixing cyanobacterium Nostoc muscorum.

Pl. Cell. Physiol. 25 : 845-850.

The nitrogenase and nitrate reductase activities of the molybdenum requiring parent Nostoc muscorum and its tungsten - and chromium requiring mutant strains, growing with Mo, W and Sr, respectively, were significantly enhanced by the addition of 0.063 uM vanadium. Such interactions were not observed with NO_2^- as a nitrogen source.

1031.
VARADE, P.B. and POKHRIYAL, T.C. (1985) Effect of nitrate on nitrogenase and nitrate reductase activities of mung bean (Vigna radiata L.). Pl Physiol. Biochem. 12 : 57-63.

Leguminous plants like mungbean contain active nitrate reductase in root as well as in root nodules. Different concentrations of nitrate were applied to the mung bean plants after one month of plantation. The plants with 5 mM nitrate treatment could maintain higher nitrogenase activity during the period of flowering and pod filling than the control plants. Root nitrate reductase activity increased in treatments that decreased nitrogenase activity. The results suggest that nodule nitrate reductase and root nitrate reductase activities can be exploited for the better yield with proper dose of fertilizers.

1032.
VENKATARAMANAN, S., NATARAJAN, K. and KANNAIYAN, S. (1986) Influence of superphosphate, Mussoorie rock phosphate, Sagar rock phosphate and pyrite on the growth of Azolla. Proc. natn. Sem. Rock-Phosphate in Agriculture, PPCL-TNAU, Coimbatore, pp 178-183.

The influence of some essential nutrients on the growth of Azolla was studied. In the field experiment the application of superphosphate at 10kg/ha level recorded the maximum growth over control (85.71%). When pyrite was added along with MRP and SRP, further increase of about 50% and 57% respectively was recorded. In the pot experiment also, application of P recorded maximum growth. Further, the addition of pyrite to P and MRP increased the growth significantly. The increase in the levels of P, MRP and pyrite gave increased growth of Azolla, showing a positive influence.

1033.
VIRAJ BERI, MEELU, O.P. and BALDEV RAJ (1984) Studies on multiplication, nitrogen fixation and utilization of Azolla. Bull. Indian Soc. Soil Sci. 13 : 387-391.

Experiments on multiplication, nitrogen fixation and utilization of Azolla in north India revealed that high temperature during summer and low temperature in winter were the limiting factors for the growth of Azolla. However, it produced 55 tonnes of green matter and 139 kg N/ha per year, mostly in the months of March, April and October. The nitrogen fixation by acetylene-reduction method indicated the presence of nitrogenase enzyme in Azolla, showing thereby that Azolla was not simply recycling the soil and water nitrogen but was capable of fixing dinitrogen from the atmosphere. However, this method estimated higher nitrogen fixation than conventional Kjeldahl method. Preliminary studies revealed that nitrogen supplied through Azolla was almost as efficient as inorganic fertilizer nitrogen.

1034.
VORA, M.S.; SHETH, R.D. and PATEL, J.P. (1985) Application of Azolla pinnata in rice cultivation. Curr. Sci. 54 : 437.

In the treatment of 75 kg N/ha, the highest grain yield of 7013 kg/ha was recorded as against 6166 kg/ha in Azolla inoculated treatment which indicated the superiority of chemical N over Azolla.

Author's Index

(Numbers refer to abstract numbers in the book)

Source Index

(Numbers refer to abstract numbers in the book)

Il Riso 30-1, 328, 521
Indian Agric. 79, 547
Indian Agric. News Digest 293

Subject Index

(Numbers refer to abstract numbers in the book)